# Tourism Geography

Tourism has become one of the most significant forces for change in the world today. Regarded by many as the world's largest industry, tourism prompts regular mass migrations of people, exploitation of resources, processes of development and inevitable repercussions on places, economies, societies and environments. It is a phenomenon that increasingly demands attention.

*Tourism Geography* reveals how geographic perspectives can inform and illuminate the study of tourism. The book explores the factors that have encouraged the development of both domestic and international forms of tourism, highlighting ways in which patterns of tourism have evolved and continue to evolve. The differing economic, environmental and socio-cultural impacts that tourism may exert upon destinations are examined, together with a consideration of ways in which planning for tourism can assist in the regulation of development and produce sustainable forms of tourism.

Drawing on case studies from across the world, *Tourism Geography* offers a concise review of established geographies of tourism and shows how new patterns in the production and consumption of tourist places are fashioning the new tourism geographies of the twenty-first century.

**Stephen Williams** is Principal Lecturer in Geography at Staffordshire University, UK.

# Routledge Contemporary Human Geography Series

Series Editors:
**David Bell** and **Stephen Wyn Williams**, Staffordshire University

This series of 12 texts offers stimulating introductions to the core subdisciplines of human geography. Building between 'traditional' approaches to subdisciplinary studies and contemporary treatments of these same issues, these concise introductions respond particularly to the new demands of modular courses. Uniformly designed, with a focus on student-friendly features, these books will form a coherent series which is up to date and reliable.

*Forthcoming Titles*:

**Urban Geography**

**Rural Geography**

**Political Geography**

**Historical Geography**

**Cultural Geography**

**Theory and Philosophy**

**Development Geography**

**Transport, Communications & Technology Geography**

**Routledge Contemporary Human Geography**

# Tourism Geography

## Stephen Williams

**Routledge**
Taylor & Francis Group

LONDON AND NEW YORK

First published 1998
by Routledge
11 New Fetter Lane, London EC4P 4EE

Simultaneously published in the USA and Canada
by Routledge
29 West 35th Street, New York, NY 10001

Reprinted 2000, 2001, 2002 (twice), 2003, 2004

*Routledge is an imprint of the Taylor & Francis Group*

© 1998 Stephen Williams

The right of Stephen Williams to be identified as the Author of this Work
has been asserted by him in accordance with the Copyright, Designs and
Patents Act 1988

Typeset in Times by Keystroke, Jacaranda Lodge, Wolverhampton
Printed and bound in Great Britain by St Edmundsbury Press Limited,
Bury St Edmunds, Suffolk

*British Library Cataloguing in Publication Data*
A catalogue record for this book is available from the British Library

*Library of Congress Cataloging in Publication Data*
Williams, Stephen, 1951–
    Tourism geography / Stephen Williams.
    p.   cm. – (Routledge contemporary human geography series)
    Includes bibliographical references and index.
    1. Tourist trade. I. Title. II. Series.
G155.A1W49   1998
    338.4′791–dc21
                                                    98-6809
                                                    CIP

ISBN 0–415–14214–8 (hbk)
ISBN 0–415–14215–6 (pbk)

# Contents

# Figures

# Tables

# Boxed case studies

# Acknowledgements

I am pleased to acknowledge the assistance and support of a number of individuals and organisations that have made the production of this volume possible. The active interest and encouragement of the Series Editors – David Bell and Stephen Wyn Williams (who are both colleagues at Staffordshire University) – together with Sarah Lloyd and her staff at Routledge, were instrumental in seeing the work through to fruition and I am grateful for their considerate and patient attention.

I must extend particular thanks to Rosemary Duncan (Cartographer in the Division of Geography, Staffordshire University), who transformed my rough-and-ready sketches into maps and illustrations that are both a genuine embellishment and an integral part of this book. Her cheerful willingness to rework ideas that didn't succeed first time, her careful attention to detail and the skills that she devoted to the work are greatly appreciated. Thank you, Rosie!

Writing a book is a lengthy and solitary process but the long hours were often enhanced by the occasional company of J.S., R.V.W. and E.J.M. Over the years, these have become old friends and I cannot now imagine working without them.

Lastly, I am grateful to the following for permission to reproduce material under copyright: the Canadian Association of Geographers (Figure 2:4); Blackwell Publishers Ltd (Figure 4:8); Elsevier Science Ltd (Figures 4:4; 4:5 and 6:4); John Wiley & Sons Inc. (Figure 6:2) and Routledge Ltd (Figure 7:2).

# 1 ▶ Issues and approaches in the contemporary geography of tourism

Thirty years ago, the inclusion of a book on tourism within a series of introductory texts covering differing aspects of human geography would have been an unlikely event. Today, the exclusion of tourism from the geography curriculum seems equally improbable. From a position at the end of the Second World War when relatively few people travelled for the purposes by which we now define the activity, tourism has grown to a point at which it is commonly being heralded as the world's largest industry. The World Tourism Organization (WTO) estimates that international travellers today number in excess of 528 million people annually with yearly gross receipts from their activities exceeding US$320 billion. To these foreign travellers and their expenditure must be added the domestic tourists who do not cross international boundaries but who, in most developed nations at least, are several times more numerous than their international counterparts. (In the UK holiday tourism sector, for example, for each foreign visitor there are around four domestic holidaymakers and significantly more day trippers.) Globally, an estimated 74 million people derive direct employment from the tourism business: from travel and transportation, accommodation, promotion, entertainment, visitor attractions and tourist retailing. Tourism has been variously advocated as a means of advancing wider international integration within areas such as the European Union (EU) or as a catalyst for modernisation, economic development and prosperity in emerging nations in the Third World.

Yet tourism also has its negative dimensions. Whilst it brings development, tourism may also be responsible for a range of detrimental impacts on the physical environment: pollution of air and water, traffic congestion, physical erosion of sites, disruption of habitats and the species that occupy places that visitors use, and the unsightly visual blight that results from poorly planned or poorly designed buildings. The exposure of local societies and their customs to tourists can be a means of sustaining traditions and rituals, but it may also be a potent agency for cultural change, a key element in the erosion of distinctive beliefs, values and practices and a producer of nondescript, globalised forms of culture. Likewise, in the field of economic impacts, although tourism has shown itself to be capable of generating significant volumes of employment at national, regional and local levels, the uncertainties that surround a market that is more prone than most to the whims of fashion can make tourism an insecure foundation on which to build national economic growth, and the quality of jobs created within this sector (as defined by their permanence, reward and remuneration levels) often leaves much to be desired.

Readers will detect within this medley of themes and issues much that is of direct interest to the geographer, and to disregard what has become a primary area of physical, social, cultural and economic development would be to deny a pervasive and powerful force for change in the world in which we live.

## What is tourism?

But what is tourism and how does it relate to associated concepts of recreation and leisure? The word 'tourism', although accepted and recognised in common parlance, is nevertheless a term that is subject to a diversity of meanings and interpretations. For the student this is a potential difficulty since consensus in the understanding of the term and, hence, the scope for investigation that such agreement opens up, is fundamental to any structured form of enquiry and interpretation. Definitional problems arise partly because the word 'tourism' is typically used as a single term to designate a variety of concepts, partly because it is an area of study in a range of disciplines (geography, economics, business and marketing, sociology, anthropology, history and psychology), and the differing conceptual structures within these disciplines lead inevitably to contrasts in perspective and emphasis. It is

A third practical problem is the lack of a conceptual grounding for the study of tourism. This is important because in the absence of theoretical underpinning, adopted methodologies tend to regress towards a broadly empirical/descriptive approach, and the insights that can arise from the more structured forms of analysis that a sound conceptual framework can provide tend to be lost. This is not to imply that there has been no conceptualisation within the study of tourism, for (as several of the following chapters will demonstrate) the understanding of many aspects of tourism has benefited from varying degrees of theoretical thought. But what is largely absent is the broader synthesis of diverse (though still related) issues and perspectives. As an intrinsically eclectic discipline, geography is better placed than many to provide the type of holistic perspective that a multi-dimensional phenomenon such as tourism evidently merits, but there are still limits to the level and extent of understanding that any one discipline, in isolation, can afford. The student of tourism must, therefore, be predisposed to adopting multi-disciplinary perspectives in seeking to understand this most contradictory and, at times, elusive of phenomena.

This, however, creates a fourth problem. Adopting a multi-disciplinary approach is easier said than done since the application of alternative perspectives can, when confronted with a multi-faceted activity such as tourism, obscure as much as it reveals. This is because the student will encounter different explanations of tourism and tourist behaviour that may appear outwardly to be contradictory. As an example, I shall introduce briefly the key area of tourist motivations.

## Tourist motivations

The question of *why* people travel is both obvious and fundamental to any understanding of the practice of tourism and its consequences, including the geography of tourism. However, although there is general (though not universal) agreement that the primary motive for pleasure tourism is a real or perceived need to escape temporarily from the routine situations of the home, the workplace and the familiarity of their physical and social environments, the many theories of tourist motivation may differ quite substantially in their interpretation and explanations of resulting tourist patterns and behaviours. Examples of three types of motivation theories may serve to illustrate the point.

First, there are a number of theories that focus on the analysis of tourist behavioural patterns as a means of exposing tourist motives. One of the most interesting is Graburn's explanation of tourist 'inversions' – shifts in behaviour patterns away from a norm and towards a temporary opposite. This might be shown in extended periods of relaxation (as opposed to work); increased consumption of food, and increased purchases of drinks and consumer goods; relocation to contrasting places, climates or environments; or relaxation in dress codes through varying states of nudity. Graburn proposes several different headings or 'dimensions' under which tourist behavioural inversions occur, including environment, lifestyle, formality and health (Table 1:1). Graburn emphasises that within the context of any one visit, only some dimensions will normally be subject to a reversal, and this allows us to explain how the same people may take different types of holiday at different times and to different locations. It is also the case that actual behaviour patterns will usually exhibit degrees of departure from a norm, rather than automatically switching to a polar opposite. This accommodates those people whose behavioural patterns as tourists show minimal differences from most of the normal dimensions of their lives, whilst still emphasising the notions of escape and contrast as being central to most forms of tourism experience.

A second set of motivational theories place their emphasis upon the idea that tourist movements are a product of a combination of factors that both prompt the participant to leave their present location and attract them to another – a push–pull effect. This idea is implicit in many motivational theories but perhaps receives its most explicit statement in Iso-Ahola's model of the social psychology of tourism. Here elements of escape from routine environments are deliberately juxtaposed with a parallel quest for intrinsic rewards in the environments to be visited. By envisaging these key elements as the axes on a matrix (Figure 1:2) it is possible to construct a set of theoretical 'cells' in which elements of escape and reward are combined in differing ways and within which tourist motives may be located, depending upon their particular circumstances and objectives at any one time. As with Graburn's conceptualisation, Iso-Ahola's model has an overtly dynamic quality since the positioning within the matrix may change, both within the context of a tourist trip and between trips, as needs and motives change or fluctuate. Iso-Ahola places the emphasis upon social dimensions to motivation, but ideally the matrix requires a third dimension that incorporates the physical environment, since there is an abundance of

**Table 1:1** *Examples of 'inversions' in tourism*

| Dimension | Continua | Tourist behavioural pattern |
|---|---|---|
| Environment | Winter vs. summer<br>Cold vs. warmth<br>Crowds vs. isolation<br>Modern vs. ancient<br>Home vs. foreign | Tourists escape cooler latitudes in favour of warmer places. Urban people may seek the solitude of rural or remote places. Historic sites attract tourists who live in modern environments. Familiarity of the home replaced by the difference of a 'foreign' environment. |
| Lifestyle | Thrift vs. indulgence<br>Affluence vs. simplicity<br>Work vs. leisure | Expenditure increased on special events or purchases. Experiences selected to contrast routine of work with rewards of leisure. |
| Formality | Rigid vs. flexible<br>Formal vs. informal<br>Restriction vs. licence | Routines of normal time-keeping, dress codes and social behaviours replaced by contrasting patterns and practices based on flexibility and informality. |
| Health | Gluttony vs. diet<br>Stress vs. tranquillity<br>Sloth vs. exercise<br>Age vs. rejuvenation | Tourists indulge through increased consumption. Relaxation sought as relief from routine stresses. Active holidays chosen as alternative to sedentary patterns in daily lives. Health spas and exercise used to counteract processes of ageing. |

Source: Adapted from Graburn (1983).

empirical evidence suggesting that a change of place (with the possibility of attendant changes in landscape, climate, etc.) is of equal if not greater importance than socially based factors.

Then there are a third set of motivational theories that approach the task of explaining patterns of tourism by relating those patterns to personal characteristics of individuals. The most commonly used example is Plog's psychographic profile approach. Plog envisages populations as being arranged along a personality continuum. At one pole are people whom he labels 'psychocentrics' – essentially self-inhibited,

**Figure 1:2** *Iso-Ahola's model of the social psychology of tourism*

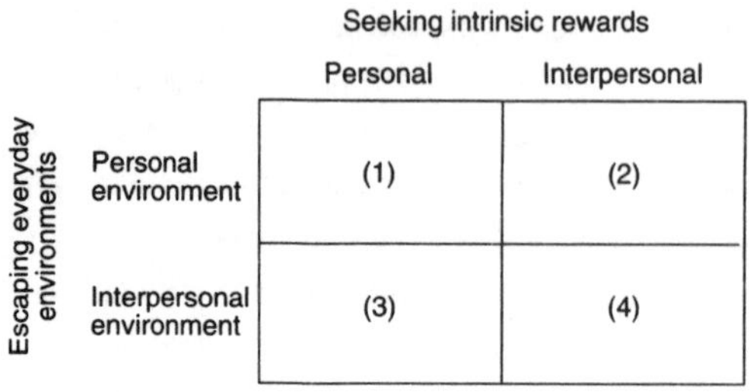

Seeking intrinsic rewards

|  | Personal | Interpersonal |
|---|---|---|
| Personal environment | (1) | (2) |
| Interpersonal environment | (3) | (4) |

Escaping everyday environments

Source: Iso-Ahola (1982).

non-adventurous types – whilst at the other he locates 'allocentrics', who display opposing personalities – confident people who are naturally adventurous and who seek variety and experience. Between these poles are arranged intermediate categories that reveal greater or lesser tendencies to allocentricism or psychocentricism (Figure 1:3a). According to Plog, personality traits may then be linked to travel characteristics, the proposition being that the psychocentrics gravitate towards the familiar destinations, are more likely to demand tourist services (such as accommodation and food) that match their normal patterns of consumption and will prefer package tours. In contrast, the allocentric tourist is more likely to act independently, to seek out novel destinations and different forms of experience. With this as a framework, Plog suggests it is possible to match destination patterns to personality and, on the basis of a study of American tourists, he annotated the basic psychographic continuum to illustrate likely destinations of the different groups (Figure 1:3b).

Plog's model has been widely criticised as over-simplifying a complex process by seeking an explanation based upon one element, and it lacks the dynamic qualities that are essential to explaining how the same individuals can alter their behaviour patterns between different tourist trips. But it remains a theory widely discussed in tourism textbooks and, in the context of this particular discussion, illustrates well how distinctly different approaches can address the same basic question.

## Tourism typologies

Reconciliation of apparently contrasting views of the type that are illustrated by discussion of tourist motivation is problematic until it is realised that one of the possible explanations is that many tendencies will exist simultaneously. One of the intractable problems of those tourist studies that seek to isolate generalities within the patterns is that the real-world complexity of tourism admits a whole spectrum of motives and

**Figure 1:3** *Plog's psychographic tourist profile*

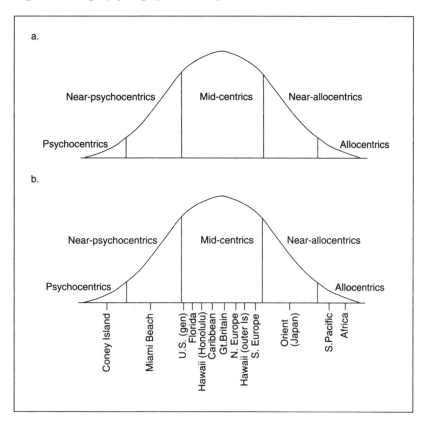

Source: Plog (1974).

behaviours that in some cases *will* co-exist not just within similar groups but within individuals, varying in impact and significance according to circumstance. Within the area of tourism as a whole, there are clearly many different types of tourist and situations under which tourism will develop. This realisation has led to a number of attempts at defining alternative structures of tourism and typologies of contrasting categories of tourist.

The benefits of recognising typologies of tourists and tourism are that they allow us to identify key dimensions of the activity and its participants. In particular, typological analyses help us to:

● recognise different types of tourism (for example, recreational or business tourism);

- recognise different types of tourist (for example, organised mass tourists, independent travellers or lone explorers);
- anticipate contrasting motives for travel;
- expect variations in impacts within host areas according to motives and forms of travel;
- expect differences in the structural elements within tourism (for example, accommodation, travel and entertainment) that different categories of tourists will generate.

Attempts at the categorisation of tourism normally use the activity that is central to the trip as a basis around which to construct a subdivision. Thus we may draw important distinctions between recreational tourism (where activities focus upon the pursuit of pleasure, whether through passive enjoyment of places as sightseers or through more active engagement with sports and pastimes) and business tourism (where the primary focus will be the development or maintenance of commercial interests or professional contacts). However, it is also recognised that people may travel to secure treatment for medical conditions, for educational reasons, for social purposes or, as pilgrims, for religious purposes. Furthermore, most of these categories may themselves be subdivided. In recreational tourism, for example, we may wish to differentiate the modern-day equivalents of the Grand Tourists (see Chapter 3) who travel to experience foreign cultures, history and heritage from others who simply laze by the poolside or who sightsee in the somewhat detached manner of the casual visitor. It is, though, risky to push such distinctions too far or to assume that tourists travel for a narrow range of reasons. Most tourists choose destinations for a diversity of purposes and will combine more than one form of experience within a visit.

There are many typologies of tourists but if we distil their essence, we may perhaps derive the following broad outline summary, wherein four types of tourist are recognised:

1 *Organised mass tourists*. These people travel to destinations that are essentially familiar rather than novel – familiarity commonly having been gained through previous experience, through reported experiences of others or through media exposure. A sense of familiarity is reinforced by the nature of goods and services available at the destination, for example the retailing of tea, English beer and fish and chips at mass tourist locations on the Spanish Mediterranean coast. The mass tourist is highly dependent upon travel industry

infrastructure to deliver a packaged trip at a competitive price and with minimal organisational requirements on the part of the tourist. Incipient tourists, feeling their way into foreign travel and destinations for the first time, may typically operate in this sector, at least until experience is acquired. This sector is dominated by recreational tourists.

2 *Individual or small-group mass tourists.* In this sector, tourists will be partly dependent upon the infrastructure of mass tourism to deliver some elements of the tourist package, especially travel and accommodation, but will structure more of the trip to suit themselves. The experiences sought are still likely to be familiar but with some elements of exploration. The sector will contain business tourists alongside recreational travellers and is also more likely to accommodate activities such as cultural or educational forms of tourism.

3 *Lone travellers and explorers.* In this form of tourism much more emphasis is placed upon the tourist's willingness and desire to arrange his or her own trips, and these people are usually seeking novelty and experiences that are not embodied in concepts of mass tourism. Hence, for example, contact with host societies will be more important. It is possible, too, that people with very specific objectives in travelling (for example, some business tourists or religious or health tourists) would be more prominent here. There is still a residual dependence upon elements in the tourism industry – travel and accommodation bookings being the most likely point of contact.

4 *Drifters.* Some authors have distinguished a fourth category of tourists who probably do not consider themselves to be tourists in any conventional sense. They plan trips alone, shun other tourist groups (except perhaps fellow drifters) and seek immersion in host cultures and systems. People engaged in this form of tourism are often pioneers, constituting the first travellers to previously untouched areas.

Understanding of these typological subdivisions of tourists is developed further when linked to contrasting patterns of tourist motivation. The actions of organised, mass tourists, for example, have been interpreted as essentially a quest for pleasure that may be *diversionary* – that is, escaping from boredom or the repetitive routine of daily life – or *restorative* – perhaps through rest, relaxation and entertainment. The individual or small-group traveller may retain all or some of these motives but might equally replace or supplement them with an *experiential* motive, a desire to learn about or engage with alternative customs or cultures. Some writers have interpreted such actions as a quest for authenticity or meaning in life of a kind that modern industrialised societies seem less and less able to furnish. This tendency becomes most clearly embodied in the motives of the explorers and the drifters who, it is argued, seek active immersion in alternative lifestyles in a search for a particular form of self-fulfilment.

Clearly, these different patterns of activity and behaviour will lead to a range of impacts (especially upon host areas) and exert particular demands in respect of structures that need to be in place. Organised mass tourism, for instance, imposes infrastructure upon host areas: extensive provision of hotels and apartments, entertainment facilities, transportation systems, public utilities, etc. that inevitably alters the physical nature of places and will probably affect environments and ecosystems too. The actions of tourists *en masse* will usually have an

Figure 1:4 *Tourism and tourists: a typological framework*

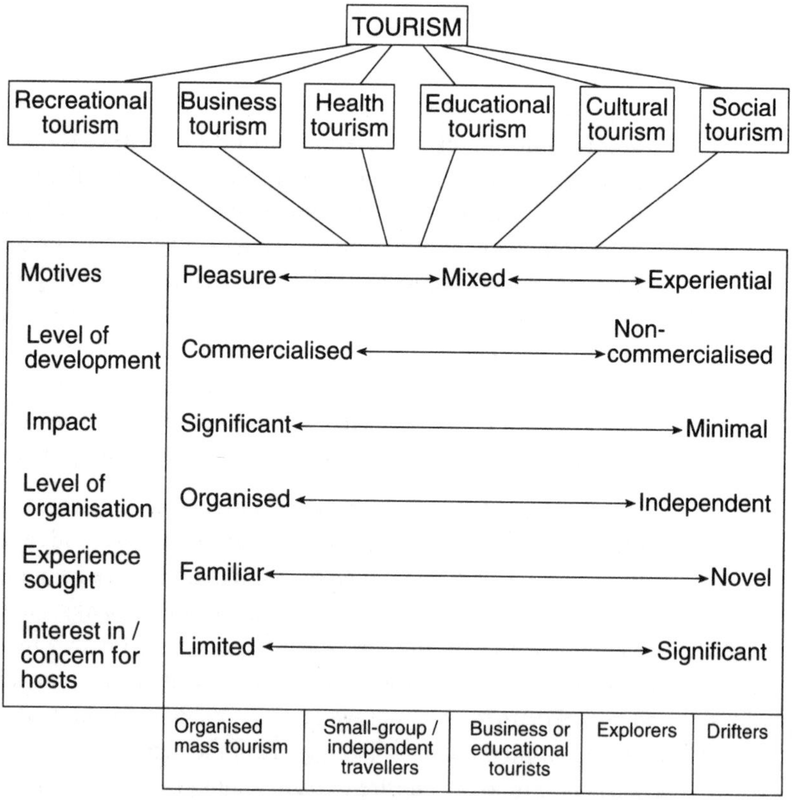

impact upon local lifestyles. The much smaller numbers of explorers, in contrast, make fewer demands for infrastructure provision, and through different attitudes and expectations towards host communities exert a much reduced impact upon local life.

These ideas are summarised in Figure 1:4, which offers a typological framework of tourism and tourists as an aid to recognising the dimensions of what is a highly segmented market or sphere of activity. It is important to appreciate in interpreting this summary, however, that as individuals we can and will move around within the framework, especially as we progress through the life cycle.

## The structure of the tourism experience

Now that we have seen, in summary form, the main elements that help us define the structure of tourism, it is also useful to consider the key structural elements in the tourism experience and how they inter-relate. This is set out in a summary diagram in Figure 1:5, which proposes that the tourism experience comprises:

- An initial phase of *planning the trip* in which destinations, modes of travel, preferred styles and levels of accommodation are evaluated and a destination selected. The planning phase is informed by a number of potential inputs, including previous experience, images and perceptions of places and suggestions made by others.
- *Outward travel*. All tourism involves travel, and it is important to realise that travelling is often more than just a means to an end. In many tourism contexts, getting there is half the fun, and in some forms of tourism – most conspicuously in sea cruising – the act of travelling rather than visiting places often becomes the central element within the tourism experience as a whole.
- *Experience at the destination*. This element is normally the main component within the visit and most clearly reflects the category or categories of tourism in which the trip is located and the motivations of the visitors. Typically experience at the destination will include elements of sightseeing, leisure shopping and the collection of souvenirs and memorabilia. It may also include varying levels of contact with host populations, society and culture, the extent and significance of which will vary.
- *Return travel*, which, as with the outward journey, may be an integral part of the tourism experience, although it may not realise the same

degree of pleasure, anticipation and excitement, as the trip is nearing its end and fatigue may have begun to affect the tourist.

● *Recall.* The trip will be relived subsequently and probably repeatedly, in conversation with friends and relatives, in holiday photographs and/or videos, or in response to the visual prompts offered by souvenirs that may now be arranged around the home. The recall phase will also inform the preliminary planning of the next visit and may be either a positive or a negative stimulus, depending upon the success or failure of the trip.

**Figure 1:5** *Structure of the tourist experience*

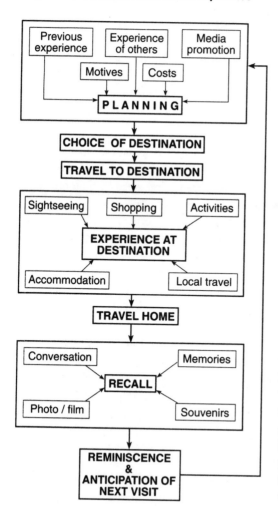

## Geography and the study of tourism

But what can geographers bring to the study of this field? Tourism (with its focus upon travelling and the transfer of people, goods and services through time and space) is essentially a geographical phenomenon, and accordingly there are a number of ways through which a geographical perspective can illuminate the subject:

*The effect of scale.* To treat tourism as if it were a phenomenon that is consistent in cause and effect through time and across space is to misrepresent the dynamic diversity that is naturally present. However, the spatial perspective allows us initially to recognise and make a valuable distinction between activity at a range of geographical scales – global, international, regional and local – and then to relate how patterns of interaction, motives for travel and its effects and impacts vary as the scale alters. Without such differentiation some significant parallels and contrasts will remain largely obscured.

*Spatial distributions of tourist phenomena.* This is a traditional area of interest for geographers and is concerned with several central elements within tourism as a whole. This includes the spatial patterning of supply, including the geography of resorts, of landscapes, places and attractions deemed of interest to tourists or locations at which activity may be pursued. Furthermore, geographers have a role to play in isolating patterns of demand and associated tourist movements. Where are the primary tourist-generating regions, how are they tied to the receiving areas by transportation networks and what are the characteristic forms of flows of visitors between generating and receiving areas?

*Tourism impacts.* Geographers also have a bona fide interest in the resulting impacts of tourism since these exhibit variations across time and space too. Impact studies have conventionally considered the relatively broad domains of environmental, economic, social and cultural impacts, each of which has a geographical dimension. Indeed, it may be argued that geographers need to be more active in exploring these issues. If we limit ourselves to conventional geographic concerns for spatial patterns of people, resources and tourism flows, we gain only a partial view of what tourism is about. Geography has the capacity to provide a synergistic framework (i.e. a combining approach that emphasises that the product is often more than the sum of individual parts) for exploring more complex issues such as the nature of links between tourism and development processes or the socio/cultural/anthropological concerns for host–visitor relationships.

*Planning for tourism.* As it has developed, tourism has inevitably become a focus of attention in spatial and economic planning. The capacity for physical development of tourism infrastructure to exert extensive changes in host areas is considerable, and in order to minimise detrimental influences and maximise the beneficial attributes of tourism, some form of planned development of the industry is often deemed essential. The historically close links between geography and planning (with their shared interests in the organisation of people, space and resources) therefore provide a fourth area in which geographers may contribute to the understanding of tourism.

*Spatial modelling of tourism development.* Finally, a fifth area of geographical interest has been identified in the broad field of spatial modelling of tourism development processes. Although (as noted above) the conceptual basis to the understanding of tourism is not as well grounded as it could be, a range of theoretical issues within which a geographical dimension may be discerned have been identified. These have included (as examples) attempts at modelling:

- evolution and change of patterns of tourism through time at a range of geographical scales;
- spatial diffusion of tourism, both within and between countries;
- development of hierarchies of resorts and tourism places;
- effects of distance on patterns of tourist movements.

From this outline summary of the geographer's potential interests in tourism, we may set out a list of key questions that should inform and shape an approach to the study of tourism within the discipline.

- Under what conditions (physical, economic, social) does tourism develop, in the sense of generating both demand for travel and a supply of tourist facilities?
- Where does tourism develop and in what form? (The question of location may be addressed at a range of geographical scales whilst the question of what is developed focuses particularly upon provision of infrastructure.)
- How is tourism developed? (This question will address not just the rate and character of tourism development but also the question of who are the developers.)
- Who are the tourists (defined in terms of their number, characteristics, travel patterns, etc.) and what are their motives?
- What is the impact of tourism upon the physical, economic and socio-cultural environments of host areas?

These questions form the primary issues that are developed in succeeding chapters in this book.

## Summary

Tourism has become an activity of global significance, and as an inherently geographical phenomenon that centres upon the movement of people, goods and services through time and space it merits the serious consideration of geographers. Our understanding of tourism is, however, complicated by problems of definition, by the diversity of forms that the activity takes, by the contrasting categories of tourists, and by the different disciplines in which tourism may be studied. Geography, as an intrinsically eclectic subject with a tradition in the synthesis of alternative perspectives, is better placed than many to make sense of the patterns and practices of tourists. The chapter identifies several areas of study in which geographical approaches can aid wider understanding of tourism, including the spatial distribution of tourism, analysis of impact, tourism planning and spatial modelling.

## Discussion questions

1 Why is the definition of 'tourism' problematic?
2 What can geographers bring to the study of tourism?
3 Explain why it is important to distinguish between different categories of tourist and forms of tourism.
4 In what ways are the boundaries between 'leisure', 'recreation' and 'tourism' becoming increasingly indistinct?

## Further reading

Problems of defining and approaching tourism form a common starting-point for texts on the subject and useful discussions may be located in:
Gilbert, D.C. (1990) 'Conceptual issues in the meaning of tourism'. In Cooper, C.P. (ed.) *Progress in Tourism, Recreation and Hospitality Management*, Vol. 2, London: Belhaven: 4–27.
Mathieson, A. and Wall, G. (1982) *Tourism: Economic, Physical and Social Impacts*, Harlow: Longman.
Murphy, P.E. (1985) *Tourism: A Community Approach*, London: Routledge.
Pearce, D. (1987) *Tourism Today: A Geographical Analysis*, Harlow: Longman.
—— (1989) *Tourism Development*, Harlow: Longman.
Theobald, W. (1994) 'The context, meaning and scope of tourism'. In Theobald, W. (ed.) *Global Tourism: The Next Decade*, Oxford: Butterworth Heinemann: 3–19.

The relationship between tourism, recreation and leisure has been examined less fully, but a valuable introductory discussion is provided in:
Shaw, G. and Williams, A.M. (1994) *Critical Issues in Tourism: A Geographical Perspective*, Oxford: Blackwell: ch. 1.

Fuller explorations are provided in review articles by:
Moore, K., Cushman, G. and Simmons, D. (1995) 'Behavioural conceptualizations of tourism and leisure', *Annals of Tourism Research*, Vol. 22 No. 1: 67–75.
Smith, S.L.J. and Godbey, G.C. (1991) 'Leisure, recreation and tourism', *Annals of Tourism Research*, Vol. 18 No. 1: 85–100.

A useful discussion of tourist motivation theory is provided by:

Pearce, P.L. (1993) 'Fundamentals of tourist motivation'. In Pearce, D.G. and Butler, R.W. (eds) *Tourism Research*, London: Routledge: 113–134.

Typologies of tourism and tourists are discussed in several of the sources cited above, including Mathieson and Wall (1982), Murphy (1985) and Moore *et al.* (1995), whilst considerations of the specific contribution that geographers can bring to the study of tourism are included in Theobald (1994) and in:

Mitchell, L.S. and Murphy, P.E. (1991) 'Geography and tourism', *Annals of Tourism Research*, Vol. 18 No. 1: 57–70.

Pearce, D. (1994) 'Alternative tourism: concepts, classifications and questions'. In Smith, V.L. and Eadington, W.R. (eds) *Tourism Alternatives: Potentials and Problems in the Development of Tourism*, London: John Wiley: 15–30.

Smith, R.V. and Mitchell, L.S. (1990) 'Geography and tourism: a review of selected literature'. In Cooper, C.P. (ed.) *Progress in Tourism, Recreation and Hospitality Management*, Vol. 2, London: Belhaven: 50–66.

# 2 From stagecoach to charter plane: the popularisation of tourism

This chapter adopts a largely historical-geographic perspective with the objectives of exploring the spatial, social and structural development of domestic tourism within one country, using the case of Britain as an extended example. Britain is a particularly good case study of the development of a tourist area since it was one of the first nations to develop the practice of tourism, and it clearly exemplifies most of the factors that shape the geography of tourism. Countries that have developed tourist industries more recently than Britain will naturally reveal a more rapid and temporally compressed pattern of development, but it is contended that much of the development sequence will still be broadly comparable.

The socio-geographic development of tourism has been influenced by many elements but there are four factors that, it is argued, are especially important in understanding how and why the spatial patterns of activity have altered through time.

First, we should acknowledge the significance of change through time in attitudes and motivations. In the modern world, travel is a seemingly natural and incidental part of life and most of us harbour expectations of becoming tourists at least on an annual basis, if not more frequently. However, this was not always the case. For most of recorded history, travel was difficult, expensive, uncomfortable and often dangerous, so the desire to travel must initially have been prompted by powerful and very basic motives. It is not surprising, therefore, that amongst early tourists we find religious pilgrims motivated by a strong sense of spiritual purpose, or travellers who journeyed in the quest for health (one of the more fundamental human concerns). As travel became less difficult and more affordable, it became easier to admit other motives as a basis to tourism, especially the pursuit of pleasure. However, when differences in priorities emerge, changes in the needs, expectations and attitudes of visitors usually recast the geography of tourism and rework the character of the tourism experience.

Second, the social and economic emancipation of the urban middle classes and (particularly) the proletariat is important. For ordinary people to bring tourism into their lifestyles, extensive and fundamental change in the way in which lives were lived was required. Central to this transformation is the liberation of blocks of time that are free from work and which are sufficiently extended to permit tourism trips and, equally significantly, the ability to accumulate reserves of disposable income that can be expended on a discretionary purchase such as a holiday.

Third, mass forms of tourism became possible only with the development of efficient and affordable systems of transportation. The railway, in particular, made mass travel a reality in the second half of the nineteenth century, extending the range over which people could travel for pleasure and prompting development of new tourist regions, much in the same way that developments in civil aviation following the end of the Second World War underpinned the more recent emergence of international forms of tourism (see Chapter 3).

Fourth, modern tourism requires organisational systems and the provision of a supporting infrastructure of facilities and personnel able to run the tourism business. With the exception, perhaps, of the more solitary and explorative forms of tourism practised by the lone travellers and drifters discussed in Chapter 1, most forms of travel will not develop in the absence of the basic facilities of support. These include accommodation, transport, entertainment, retailing and (increasingly) forms of packaged tourism in which all these elements may be purchased within a single transaction. As we shall see, changes in structural elements may also be linked with changes in the geography of tourism.

## The first geography of tourism: the formation of resorts

Although tourism today is found widely across cities, countryside and coast, historically its development was concentrated into resorts, and there remains a strong, visible legacy of resort-based tourism within contemporary patterns, both in Britain and elsewhere. In Britain, the first resorts were the inland health spas – towns and villages that by chance possessed local mineral waters that were believed to have curative qualities and which attracted people who were seeking a cure for particular conditions. Mineral water cures were not an innovation (as the Roman remains at Bath bear testimony), and the intermittent and usually localised popularity of spas was a feature of the sixteenth and seventeenth centuries. However, in the mid-eighteenth century, watering places such as Buxton, Harrogate, Scarborough, Tunbridge Wells and Bath itself all enjoyed a significant increase in fortune, as mineral cures became widely popular amongst people of wealth.

Initially, of course, spas were predominantly the resort of the sick, but because they were often promoted by shrewd entrepreneurs as exclusive places – many of which also benefited from royal patronage – the spas rapidly developed as places of fashion and attracted the leisure-seeker who had no need for a cure, but was drawn by the social life that developed at the resort. In order that visitors might be entertained, facilities were provided not just for taking the waters, but for concerts and theatre, dances, walks and promenades, and at the best spas (Bath or Tunbridge Wells, for example) there soon emerged a microcosm of fashionable London life.

The geographic shift in this early form of tourism from inland spas to coastal resorts came about through an almost incidental shift in medical thinking that suggested that sea bathing and, in some cases, drinking of sea water was a more effective treatment than many of the cures offered at inland spas. Although other physicians seem to have been recommending sea water cures at broadly the same time, credit for this innovation is usually given to a Dr Richard Russell who practised near Brighton and who published an influential text on sea water cures in 1750. The idea of sea bathing rapidly caught the imagination of the upper classes (who were the only social group that could afford the time and the expense to travel to the seaside), and in a process that almost precisely mirrored the development of inland health spas, a string of fashionable and exclusive new sea bathing resorts sprang up, especially along the coasts of Kent and Sussex (see Figure 2:1), which are relatively accessible from London.

The shift from inland spas to sea bathing resorts was remarkable not only as a geographic process but also because it reflected quite profound changes in public attitudes towards the coastline. From late twentieth-century perspectives the attraction of the sea seems entirely natural, but historically the sea and its coastlines were viewed quite differently. The coast was often a place of fear and repulsion. It was a zone of tension, associated with pirates and smugglers, shipwrecks and places of invasion, whilst the sea itself was an unfathomable mystery, a home to monstrous creatures and a chaotic remnant of the Great Flood. As if to reinforce the point, the incidence of sea-sickness amongst early tourists who did venture onto the oceans must have confirmed for many that this was not a natural and proper place for people.

Yet by the beginning of the nineteenth century, the sea and the coast had become central to the popular imagination – a reflection of several changes in attitudes that included the new popularity and influence of natural theology (in which the enjoyment of natural spactacles such as the sea was now seen as a celebration of God's work); interest in 'new' sciences such as geology and natural history that focused attention upon coastlines as field laboratories; and the emergence of a public taste for the picturesque in the latter half of the eighteenth century and of the romantic movement of the early nineteenth century.

## The popularisation of the seaside

However, although the seaside might have become a focal point for popular interest, it remained relatively inaccessible in both a spatial and a social sense. In an era when roads were bad and travel expensive, the numbers who travelled to the seaside, whether for a cure or simply for pleasure, remained small. Yet within a very few years of the turn of the nineteenth century, three key changes were to transform the nature of seaside tourism.

The first of these changes came in transportation. The invention of the steamship in the early years of the nineteenth century initially prompted the growth of new resorts on the Thames estuary and also on the Clyde in Scotland, but more important changes followed the development of passenger railways after 1830. The railways transformed the nature of tourism by shortening journey times whilst increasing dramatically the numbers that might be moved on any one journey. The main effects of the railway were thus to bring existing resorts within range of the growing urban populations, to open up new areas to development and to reduce the costs of pleasure travel, although the latter effect was delayed by the initial failure of the railways to perceive the market for low-cost tourist travel.

Allied with changes in mobility came equally significant changes in social access to travel and tourism. Although popular travel for working-class families remained inhibited by a number of obvious constraints (especially lack of time and shortage of money), industrialisation in nineteenth-century Britain spawned a new and prosperous middle class

**Figure 2:1** *Expansion of sea bathing resorts in England and Wales, 1750–1900*

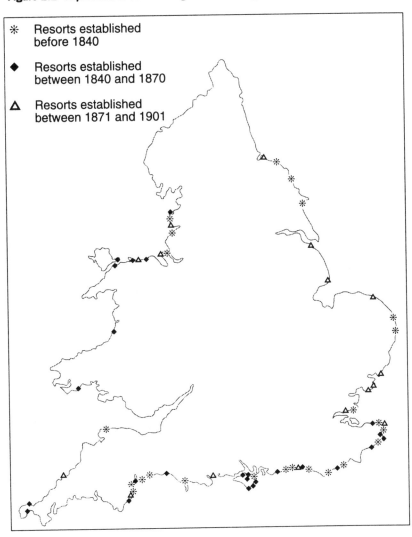

* Resorts established before 1840
♦ Resorts established between 1840 and 1870
△ Resorts established between 1871 and 1901

who were not so constrained and clearly possessed the inclination to imitate the habits of the aristocracy in resorting to the coast for day trips and holidays. The effect of this influx of new tourists on the resorts was often to displace (both spatially and temporally) the elite groups that had pioneered the resort development, and this is a tendency in tourism development that has continued to the present. However, at the same time, new spatial patterns of demand – especially from the industrial hearts of Lancashire and Yorkshire – prompted significant extensions in the numbers and locations of resorts, particularly along the north Wales, Lancashire and Yorkshire coasts (see Figure 2:1).

The third key change (or set of changes) was structural in character, embracing not just physical developments of facilities in resorts, but also the early organisation of a tourism industry. As the demand for seaside holidays grew in mid-Victorian Britain, resorts witnessed significant developments of hotels and boarding houses, places of entertainment (signalling most clearly that the motives for visiting the seaside were now largely pleasurable rather than health-related), as well as civic facilities and service industries that supported or developed around tourism. Their prosperity at this time is reflected in the fact that in the middle years of the nineteenth century, seaside resorts recorded the fastest growth rates of urban centres in Britain. (See Box 2:1 for a résumé of development processes of a typical resort – Brighton – up to 1900.)

Abilities to travel were also aided by the emergence of the first tourist excursions, the invention of Thomas Cook. Cook was a bookseller and a Baptist preacher who, on the way to a temperance meeting in 1841, had the inspired idea of chartering special trains to move supporters of temperance to meetings at low cost. However, almost immediately, the far more profitable idea of excursions for pleasure occurred to him, and by 1845 'Cook's Tours' had become a recognised phrase and the beginnings of a travel industry. Trips to the seaside were a key part of Cook's business, but thanks to the Victorian taste for the picturesque and the romantic, new tourist areas in north Wales, the Lake District, the Isle of Man, Scotland and Ireland were soon added to his itineraries, representing the first move away from a pattern of tourism centred in coastal towns.

By the last quarter of the nineteenth century, the holiday habits of the Victorian middle classes had generally begun to filter down to working people. Reductions in the length of the working week and the first

## Box 2:1

### Development of Brighton as a sea bathing resort from 1750 to 1900

In the early 1750s, Dr Richard Russell (originally a resident of Lewes, some 8 miles from Brighton) opened a practice at Brighthelmstone (as Brighton was then known) to advance his sea water treatments. On the basis of his book on sea water cures, Russell had acquired a reputation of sufficient standing to be elected a Fellow of the Royal Society in 1752, and people of wealth from all over England were attracted to his practice. Amongst these patrons were several lesser members of the royal family, including two of King George III's brothers, the Dukes of Gloucester and Cumberland, the latter of whom became a frequent visitor after 1771. The presence in the town of people of influence drew other members of fashionable society and ensured that by the time of the first visit of the Prince of Wales (later George IV) in 1783, many of the trappings of a fashionable spa were already in place, including an assembly rooms presided over by a Master of Ceremonies. The Prince of Wales continued to make annual visits to Brighton until three years before his death in 1830 and, as his stays might last anything up to three months, he arranged for the construction of the Brighton Pavilion as a summer residence. This famous building was started in 1787 although not completed until 1820.

The construction of the Pavilion was emblematic of a much wider process of physical development that was promoted by the growing fame of Brighton as a health resort, allied with the patronage of the future King. Unlike many other sea bathing places, Brighton had existed as an established town prior to its popularity as a resort. Its estimated population in 1700 is put at around 1,500, but the rise in popularity of sea bathing saw that figure increase to an estimated 3,500 in the year in which the Prince of Wales first visited, over 7,000 in 1801 and over 25,000 in 1821. (The rate of increase between 1811 and 1821 was the highest in Britain, although the expansion experienced in the new industrial cities in the north of England involved greater aggregate numbers.) In addition to this permanent population, visitor numbers grew from less than 400 in 1760 to over 7,000 in 1818. After 1815 there was significant construction of exclusive housing (some in planned developments such as at Kemp Town), as well as cheaper forms of lodging and boarding houses. Alongside house construction we also see evidence of local investment in facilities to support the resort. For example, the first pier was opened in 1823 and at the same time there was a spate of new hotel construction along the sea frontage. An emerging municipal and commercial status was also evident. The first bank dated from 1787, a new town hall and market were started in 1827, a local police force was formed in 1830, and from 1832 the town had two sitting Members of Parliament. The construction of a significant number of new churches in the town during the 1820s and 1830s is also indicative of a maturing and developed community.

The growing importance of Brighton as a resort was assisted by improvements to transportation links. The enterprise that was demonstrated by the coach operators in developing and improving services to Brighton in the first decades of the nineteenth century was remarkable, given the generally poor state of turnpike roads at the time. Yet in 1822 no less than 62 coaches ran daily to and from Brighton, including 39 to and from London, and at peak times as many as 500 people might arrive in a single day. Journey times from London had been reduced from around 9 hours in 1790 to 5 hours in 1833. However, much was to change after the arrival of the railway, which reached the town in 1841.

The primary effect of the railway was that it made the seaside accessible to the population at large – in the particular case of Brighton, accessible to the middle- and working-class populations of London. As early as Easter Monday 1844, a single excursion train from London to Brighton carried over 1,100 trippers, but this was only a hint of what would follow. For example, in one week in 1850, over 73,000 visitors reached Brighton by rail, part of a total influx for that year estimated in excess of a quarter of a million people.

The accessibility of rail travel ensured not only that visitors of all kinds began descending on Brighton, but also that the attractions of the coast as a place of residence, and the employment opportunities that seaside towns afforded, drew a growing resident population too. Thus, the permanent population which was already increasing rapidly surged forward under these new stimuli to growth. Many of Brighton's new residents were people in the working and serving classes, which even then were an important adjunct to the developing tourism industry. In 1841 the town's population was around 46,000, but ten years after the railway came to Brighton, it had increased by over 41 per cent to 65,000. Between that date and the end of the century, the population almost doubled to 123,000 people. Of course, not all these people were associated with the holiday industry. Brighton was a centre of industry in its own right with a large railway works, and retailing and commerce also provided employment for many residents. Retirement to the seaside was also becoming fashionable. Nevertheless, the emergence of a modern tourist industry in towns such as Brighton in the latter part of the nineteenth century was an important facet of the overall growth in population of these places.

The popularisation of the seaside that cheap rail travel made possible had a number of effects upon the character and social tone of the resort. After the death of George IV, both William IV and then Queen Victoria continued the royal tradition of visiting Brighton. However, significantly, Victoria's last visit was in 1845, as she found the curious gaze of the common masses (who followed her every public move) more than she could bear. Likewise, the elite visitors who continued to go to Brighton chose instead to move the fashionable season to the autumn and, eventually, the winter, to avoid the trippers who poured out of London in the summer. Contemporary accounts tell how in the winter months the promenades, piers, theatres and other entertainment continued to enjoy plenty of usage, albeit by a group that was distinctly different from the summer visitor.

However, those amongst the railway tourists who could afford to stay were sufficiently numerous to prompt another wave of hotel construction, including (amongst several large projects) the Grand Hotel (1864), which was one of the most stylish and modern hotels in the country and one of the first to employ both electric lighting and lifts to the upper floors. There was also significant investment in other visitor facilities, including (for their entertainment) the new West Pier (opened in 1868) and the Brighton Aquarium (1871), and (for their spiritual needs) several additional places of worship. The new pier, together with most of the established theatres, displayed many of the attractions that were particularly associated with the later Victorian seaside. These included band concerts (especially in the style of German military bands), music-hall acts, 'nigger minstrels' and later pierrots, many of which reflected the tastes and preferences of the working classes who, by 1900, had largely displaced the higher social classes from the forms of popular seaside that Brighton had come to represent.

Sources: Gilbert (1975), Pimlott (1947), Walvin (1978).

statutory holidays that followed Lubbock's Bank Holidays Act of 1871 had made more time available for seaside excursions, whilst by the 1880s and 1890s gradual improvements in levels of pay, when combined with the Victorian virtue of thrift, which had often been essential to basic survival in the early phases of urban industrialisation, were paying dividends in terms of the abilities of many working families to save money for excursions and holidays. In industrial communities, especially in the north of England for example, saving through co-operative or friendly societies was actively encouraged, and the benefits became manifest in a number of ways, including the taking of holidays.

The First World War marked a watershed in many aspects of life, none more so than the incidence and practice of popular forms of leisure. Whatever vestiges of exclusivity in the traditional resorts that may have survived the onslaught of middle- and working-class tourists in the nineteenth century were largely swept away by the collapse across Europe of the old orders, and as the social elites finally deserted the old resorts in favour of new and exclusive (foreign) places, as yet untouched by popular demands, the British seaside became the resort of the commoner.

The processes of social emancipation in access to tourism that had been gaining momentum at the end of the nineteenth and in the early years of the twentieth centuries continued after 1918, reinforced by:

● further advances in the incidence of paid holidays (although these were not required by law until 1938);

- new and cheaper forms of transportation (such as buses);
- active promotion of holiday regions – especially by railway companies;
- structural changes (such as new forms of low-cost holiday accommodation: holiday camps, camp sites and eventually caravans).

On the eve of the Second World War, it is estimated that some 11 million holidays and an uncounted number of day excursions were taken within Britain, the vast majority of which were directed at the seaside resorts that, by then, had become synonymous with British domestic tourism. In 1938, for example, Blackpool alone received an estimated 7 million visitors.

## Patterns of British tourism since 1945

The early post-1945 pattern of domestic tourism in Britain essentially sustained the emphasis upon coastal resorts but several significant changes quickly emerged to redraw the map of post-war tourism. Three themes are worth attention: first, the overall growth in the market; second, the spread of tourism to new areas in the countryside and eventually into major cities; and third, the stagnation and, in some circumstances, decline of traditional resorts, a consequence both of the restructuring of domestic tourism and, especially, the increase in foreign travel.

## The growth of tourism after 1945

In 1951, when the first of what was to become an annual survey of holidaymaking was conducted by the British Travel Association (later to be redesignated the British Tourist Authority, both hereafter referred to as the BTA), an estimated 26.5 million holidays were taken by the British, including 1.5 million abroad. Figure 2:2 charts the expansion in total holidays taken between 1951 and 1970 and reveals a distinct pattern with pronounced growth throughout the 1950s, followed by a period of relative stability in the 1960s. Explanation for the growth would need to take account of several factors:

- The latent demand that had built up in the latter part of the 1930s and during the war years was finally released as the Holidays with Pay Act 1938 came fully into force.

- Real wages increased bringing improved living conditions and more widespread household purchases of luxury items, including holidays.
- Holidays were actively promoted within the media, by transport operators and a rapidly developing travel industry.
- Popular expectations were that an annual holiday was now an attainable and routine part of most lifestyles.

**Figure 2:2 *Pattern of British domestic and foreign holidays, 1951–70***

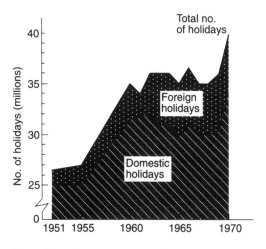

Source: British Travel Association (1969); British Tourist Authority (1995).

In comparison with the 1950s, the 1960s were a time of relative stability until a second (short) phase of growth became established around 1969. Figure 2:3 shows that growth continued up to the mid-1970s, after which a persistent decline in domestic holidays becomes an established feature as numbers of foreign holidays increase significantly. However, it is important to emphasise that these data represent growth in holidays, not necessarily growth in the numbers of holidaymakers. Indeed, after the initial expansion in numbers during the 1950s, which certainly did reflect a situation in which more people were taking a holiday, there is clear evidence that from the mid-1960s onwards much of the apparent growth in domestic tourism was solely accounted for by the increased incidence of people taking more than one holiday. Table 2:1 provides selected data from 1971 to 1993 and shows that the proportion taking a holiday scarcely changed but the numbers who took multiple holidays (and who would therefore be counted more than once in survey statistics) rose noticeably. This trend is also reflected in a 'flattening' of the holiday season, with a less pronounced peak in the traditional holiday months of July and August and greater activity in early and late summer (Table 2:2). This reflects the growing habit of taking foreign holidays in high summer and shorter breaks closer to home at other times.

Within the overall patterns of growth in tourism in the post-1945 period, there have been some significant changes in the geography of tourism in Britain. Long-term analyses of regional shifts in British tourism are

**Figure 2:3** *Pattern of British domestic and foreign holidays, 1970–93*

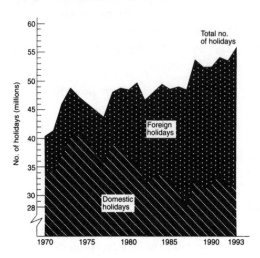

Source: British Tourist Authority (1995).

frustrated by the frequency with which the BTA redefines its regions and data areas. However, even allowing for the uncertainties that this practice creates, we may be confident that there has been a pronounced development of tourism in the South West of England (and to a lesser extent in Wales) and relative stagnation and even decline in the older holiday regions such as the North West and the South East of England (which include traditional resorts such as Blackpool, Brighton and Eastbourne). Table 2:3 illustrates estimated regional shares of the domestic market for a selection of regions that are broadly consistent in definition at a range of dates and shows the extent to which the South West now dominates the British market.

These regional shifts in domestic tourism reveal, once again, the impact of transport technology. The comparative remoteness (especially of Cornwall) had ensured that in the nineteenth century, the South West had not been extensively developed as a tourist destination, although Devon did possess some established resorts of regional importance. However, tourism to Devon and Cornwall developed substantially from the turn of the twentieth century onwards, especially in response to the active

**Table 2:1** *Level and frequency of holidaymaking (four or more nights away), 1971–93*

|  | 1971 | 1975 | 1980 | 1985 | 1990 | 1993 |
|---|---|---|---|---|---|---|
| Taking 1 holiday | 44 | 44 | 43 | 37 | 36 | 36 |
| Taking 2 holidays | 12 | 14 | 14 | 14 | 15 | 16 |
| Taking 3+ holidays | 3 | 4 | 5 | 6 | 7 | 9 |
| Total taking holidays | 59 | 62 | 62 | 57 | 58 | 61 |
| No holiday taken | 41 | 38 | 38 | 43 | 42 | 39 |

Source: British Tourist Authority (1995).
Note: All figures are percentages.

**Table 2:2** *Extension of the main holiday season, 1951–91*

| Month of main holiday | 1951 | 1961 | 1971 | 1981 | 1991 |
|---|---|---|---|---|---|
| May | 4 | 3 | 5 | 9 | 10 |
| June | 17 | 15 | 14 | 13 | 12 |
| July | 32 | 37 | 33 | 19 | 18 |
| August | 32 | 28 | 31 | 25 | 23 |
| September | 11 | 11 | 12 | 14 | 14 |
| Other month | 4 | 5 | 4 | 19 | 23 |

Source: British Travel Association (1969); British Tourist Authority (1995).
Note: All figures are percentages.

**Table 2:3** *Changes in the regional share of domestic tourism markets, 1958–93*

| | Percentage share of market | | | |
|---|---|---|---|---|
| Region | 1958 | 1968 | 1978 | 1993 |
| South West England | 14 | 23 | 20 | 25 |
| North West England | 13 | 11 | 7 | 6 |
| Wales | 10 | 13 | 14 | 12 |
| South East England | 9 | 10 | 8 | 6 |

Source: British Travel Association (1969); British Tourist Authority (1995).

**Table 2:4** *Changes in the share of holiday transport markets amongst primary modes: British domestic market, 1951–91*

| | Percentage share of market | | | | |
|---|---|---|---|---|---|
| Mode of transport | 1951 | 1961 | 1971 | 1981 | 1991 |
| Car | 28 | 49 | 63 | 72 | 76 |
| Train | 48 | 28 | 10 | 12 | 10 |
| Bus or coach | 24 | 23 | 17 | 12 | 8 |

Source: British Travel Association (1969); British Tourist Authority (1995).

promotion of the Great Western Railway, which invented the image of an 'English Riviera' for this region (see Chapter 8), whilst from about 1960 onwards, rapid increases in car ownership and the spatial flexibility that the car permits have allowed widespread diffusion of tourism, not only across the South West, but into other peripheral localities too. The shift in holiday transport from train and bus to the car has been one of the most persistent changes in the structure of tourism in Britain (Table 2:4) and has directly promoted many new tourist localities, as well as one of the most popular tourist pastimes: recreational motoring.

## New tourist areas

Although many of the holidays taken in areas such as Devon and Cornwall still retain the traditional links between holidays and the seaside, the flexibility of road travel has assisted the development of other types of tourist areas, especially in the countryside and, more recently, major cities.

Tourist enjoyment of the countryside is not, of course, a new phenomenon, and it was noted earlier how some of the first excursions of Thomas Cook took trippers to country areas rather than the seaside. As soon as the railways penetrated areas of attractive countryside, such as the Lake District or Snowdonia and areas of mid-Wales, the Victorian excursionist rapidly followed. Even before the First World War, bicycles (which appeared in number from the 1890s) opened up extensive areas of both coast and countryside to affordable forms of exploration. Hill-walking, for example on Exmoor and Dartmoor, was also becoming popular. After the First World War, bus travel began to make an impact too. In the inter-war years, coach (or charabanc) trips, as a distinctly working-class form of holiday, enjoyed widespread popularity as large numbers of people made excursions to coast and countryside, often in organised groups from churches, factories or neighbourhoods. Rambling, camping and youth hostelling developed too, as rural tourism began to emerge strongly. After 1945, the designation of ten national parks (between 1951 and 1957) provided further impetus to the development of rural tourism in England and Wales, whilst more recently the designation of country parks (since 1968) and the development of a growing range of rural tourist attractions (country houses and gardens, wildlife parks, working farms, craft centres, rural museums and steam railways) have reinforced the process.

The development of rural tourism has altered significantly the relationship between traditional seaside resorts and their hinterlands. Where once the resort would have provided the sole attraction to most holidaymakers, today the pattern is commonly one in which the resort acts as a base for widespread exploration of (rural) hinterlands, within which secondary resorts may themselves be actively developing as competing destinations. In north-east Yorkshire, for example, the popularisation of the North York Moors National Park as a tourist area for both day visitors and staying holidaymakers has shifted the balance between coastal and inland tourism quite noticeably. Historically, the resorts dominated tourism in the area, but latest estimates suggest that current levels of visits to the park are now running at over 12.7 million per year. It is true that this figure is inflated by the inclusion of people who are in transit to other tourist destinations (principally the coastal resorts of Scarborough and Whitby), but it is still significantly greater than the 3.7 million visitor days recorded for Scarborough itself. Furthermore, trends suggest that whilst the attractions within the rural areas of the national park are increasing in popularity, many of those in Scarborough and Whitby are showing declining levels of attendance. Between 1986 and 1989, the top ten attractions inside the national park increased their aggregate attendances by 14 per cent, whilst over a similar four-year period in the 1990s, most of the top attractions in Scarborough and Whitby suffered reduced attendances of the order of −3.4 per cent. At the same time, average lengths of stay in the resorts fell to just 2.6 days, a clear indication of the relative collapse of conventional patterns of staying holidays in resorts. In contrast, in neighbouring rural districts, average stays were of 4.9 days.

Traditional resorts have also lost parts of their market to competition from major cities. Historically, tourism was generally about escape from the confines of towns and cities but in one of the many reversals in convention that have accompanied the onset of a post-industrial pattern in the late twentieth century, cities themselves have now become major tourist attractions. Of course, some cities (for example, London, Edinburgh and York) have enjoyed a flourishing tourist industry for many years, typically based around sightseeing at places of interest, visiting galleries and museums, theatres and concerts, restaurants and clubs, and involving substantial numbers of foreign visitors as well as domestic tourists. What is new is the manner in which cities where there was no tradition of tourism have, through shrewd promotion and active development of attractions, been able to develop tourist industries of their own. This form of tourism typically centres upon a different set of resources from those encountered in seaside resorts – for example, leisure shopping or the enjoyment of historic and industrial

**Table 2:5** *Visitor levels to urban heritage attractions, 1993 (excl. London)*

| Attraction/location | Visitor numbers |
| --- | --- |
| Albert Dock, Liverpool | 3,500,000 [a, b] |
| York Minster | 2,250,000 [a] |
| Edinburgh Castle | 1,049,693 |
| Roman Baths and Pump Room, Bath | 898,142 |
| National Museum of Photography, Bradford | 853,784 |
| Glasgow Art Gallery and Museum | 796,380 |
| Jorvik Viking Centre, York | 752,586 |
| Wigan Pier Heritage Centre | 500,000 [a] |
| Beamish Industrial Museum, Gateshead | 400,000 [a] |
| Ironbridge Gorge Museums of Industry | 297,000 [c] |

Source: O'Brien (1990); British Tourist Authority (1995); Ironbridge Gorge Museum Trust (1996).
Notes:
[a] Estimated or rounded figure.
[b] 1989 figure.
[c] 1995 figure.

heritage – and is more important in the short-break/off-season sector of the market than for long-stay holidays. But even so, the numbers of visitors involved and the rapidity with which this sector has expanded is striking. Table 2:5 provides examples of recent levels of visiting to a cross-section of urban tourism attractions outside London.

## The decline of traditional resorts

The reshaping of the geography of tourism that is implicit in the recent development of rural and urban forms of the activity has, however, posed a major challenge to the competitive position of the traditional seaside resorts, many of which now face an uncertain future. Some writers have seen stagnation and decline as an inevitable (and natural) consequence of resort development. Figure 2:4 illustrates one conceptualisation of a resort area life cycle as developed by the geographer Butler. Most British resorts now find themselves in the critical area of the model in which, having consolidated their position in the boom years of the 1950s and early 1960s, they now have to confront the uncertainties of the alternative pathways that Butler envisages as following the 'stagnation' phase, as both spatial and behavioural patterns of tourism show visible signs of change.

**Figure 2:4** *The Butler model of resort development*

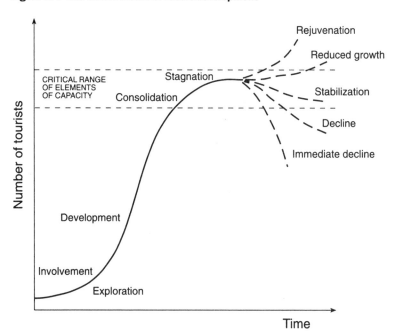

Source: Reprinted, with permission, from *The Canadian Geographer*, Vol. 24, Issue 1, 1980, article by R.W. Butler.

The decline of the domestic holiday market (which has seen levels in the mid-1990s no higher than those of 1960) has had a range of impacts upon the conventional seaside resorts, but in many places a number of distinctive and often serious problems have arisen:

- There has been a loss of traditional markets. English Tourist Board estimates suggest that by 1990, seaside resorts were attracting less than a quarter of all tourist trips, whereas in 1970 over one-third of trips had been to coastal resorts. Long-stay visitors have also been widely replaced by day trippers and touring visitors, resulting in extensive closures of (particularly) smaller hotels and guest houses.
- There has been a movement down-market. Seventy per cent of visitors to British seaside resorts are now believed to be drawn from the elderly and/or the less affluent C, D and E socio-economic groups.
- Low spending patterns by these visitors have set in motion a downward spiral of loss of income, reduced investment, diminished attraction and loss of image.

- There has been a failure to adapt and compete with new destinations, both within the UK and overseas. Alongside the newer attractions in rural and urban tourism discussed above, new concept holiday centres (such as Center Parcs) are also providing an additional source of competition.

The net effect of these changes has commonly been to impose a process of restructuring on resorts, and whilst some of the more vigorously competitive places (e.g. Brighton or Torquay) have been able to attract new investment for refurbishment and redevelopment (e.g. in stylish new shopping malls, leisure centres, conference facilities and marinas), lesser places have often had to alter significantly the facilities that they offer. Processes of adjustment in resorts have commonly seen:

- contraction in long-stay holiday provision and increased emphasis upon short-break/off-season markets;
- promotion of business and conference tourism;
- movement into 'pseudo-resort' functions, i.e. roles that benefit from traditional perceptions of the attractiveness of resorts but are not actually tourism-related, for example the conversion of small hotels and guest houses into retirement and nursing homes or office space.

Scarborough, for example, saw a 55 per cent reduction in its tourist bed spaces between 1978 and 1994, with many establishments that originally provided tourist accommodation (especially in the small hotel and guest house sector) being converted into nursing and retirement homes, flats, offices and hostels for people on state benefits. Day trips and short breaks to Scarborough District have increased, but in 1994 nearly half of these visits were directed at rural parts of the district rather than the town itself. Over the same 1978–94 period, business tourism increased significantly, with a 185 per cent growth in the number of conferences hosted in the town, but it is highly unlikely that the growth of the conference trade will be capable of compensating for the removal of other forms of tourism to alternative destinations, either within the domestic travel area or, increasingly, outside the UK.

# Summary

This chapter has presented a highly condensed account of an extensive and often complex process, but the key themes within which there is a strong geographic dimension are:

- Initially (before 1800), tourism development was highly concentrated in a small number of favoured places, access to which was generally quite localised and limited to people of wealth.
- From about 1815 onwards, as transport conditions improved and real costs of taking holidays fell, there was a marked process of spatial expansion. Original resorts were enlarged and new resorts established in response to growing popular demand, initially from the new middle classes and, in time, from working-class populations too.
- However, as resorts developed, they also tended to move down-market, as affluent visitors who formed the original clientele were displaced and became pioneers of new destinations further afield.
- After 1918, advances in road transport reinforced the accessibility and popularity of coastal resorts, but also permitted wider exploration of hinterlands, so although urban seaside resorts remained massively popular until at least 1960, rural tourism and other visitor attractions began to develop at the same time.
- After 1945, and especially after 1960, alternative destinations multiply – in rural tourism, in new forms of urban tourism and, most notably, in holidays abroad. These developments have had the effect of depressing demand for holidays in conventional resorts and prompted a realignment in the form, character and function of older seaside places.

## Discussion questions

1  What have been the primary effects of changes in transport technology on the geography of tourism in Britain?
2  To what extent may successive geographies of tourism be seen as responses to changes in social attitudes and expectations?
3  How well does the Butler model of resort development describe the evolution of British seaside resorts since 1750?
4  Taking as an example a seaside resort with which you are familiar, what evidence do you find of actions or policies directed at meeting competition from new tourism places?

## Further reading

The development of tourism from the Grand Tour to the end of the Second World War is most ably analysed in:
Pimlott, J.A.R. (1947) *The Englishman's Holiday: A Social History*, London: Faber.

whilst an equally readable account that takes the story into the post-1945 era is provided by:

Walvin, J. (1978) *Beside the Seaside: A Social History of the Popular Seaside*, London: Allen Lane.

Students with a particular interest in the growth of international tourism are recommended to read:

Turner, L. and Ash, J. (1975) *The Golden Hordes: International Tourism and the Pleasure Periphery*, London: Constable.

Changing attitudes to the coastline and the sea are perceptively analysed by:
Corbin, A. (1995) *The Lure of the Sea*, London: Penguin.

The contribution of the railways to resort development is discussed in:
Perkins, H. (1971) *The Age of the Railway*, Newton Abbot: David & Charles.

Valuable discussions of the manner in which popular demand for excursions and holidays developed in the industrial communities of nineteenth-century Britain are provided by:

Urry, J. (1990) *The Tourist Gaze: Leisure and Travel in Contemporary Societies*, London: Sage.

Walton, J.K. (1981) 'The demand for working-class seaside holidays in Victorian England', *Economic History Review*, Vol. 34 No. 2: 249–265.

whilst a thorough account of the development of holiday camps as working-class institutions is provided in:

Ward, C. and Hardy, D. (1986) *Goodnight Campers! The History of the British Holiday Camp*, London: Mansell.

Post-1945 developments in the patterns of British holidaymaking are described and summarised in:

British Travel Association (1969) *Patterns in British Holiday-making 1951–1968*, London: BTA.

For a convenient analysis of the changing geography of tourism within a region see:

Shaw, G. and Williams, A.M. (1991) 'From bathing hut to theme park: tourism development in south west England', *Journal of Regional and Local Studies*, Vol. 11 No. 1/2: 16–32.

A recent critique of the problems facing British seaside resorts is set out in:
Cooper, C.P. (1990) 'Resorts in decline: the management response', *Tourism Management*, Vol. 11 No. 1: 63–67.

The Butler model of resort development is discussed in many tourism texts but the original paper may be found in:

Butler, R.W. (1980) 'The concept of a tourist area cycle of evolution: implications for management of resources', *Canadian Geographer*, Vol. 24 No. 1: 5–12.

# 3 Shrinking world – expanding horizons: the changing spatial patterns of international tourism

In Chapter 2 we have examined a sequence of tourism development that established a familiar pattern of (largely) coastal, resort-based activity that was essentially pioneered in Britain but widely replicated (either in whole or in part), especially within other industrialised, Westernised countries. However, one of the distinguishing features of tourism is its fluidity across space and through time, so it is no surprise to find that the patterns that were set as recently as the 1950s and early 1960s are already being eroded by significant shifts in the location and character of tourist space. This is evident not just in the emergence of new destinations but also in the restructuring of established ones – a process that both reflects, and is a product of, the compression of time and space that modern transport and communications systems enable.

The basic theme of this chapter is the spatial expansion of tourism, and this is illustrated by particular reference to the development of international tourism. This is examined at both the subcontinental scale (particularly as tourism between the states of Europe) and at the intercontinental or global scale. Although international tourism is not a new phenomenon, the rapidity with which it has grown in the post-1945 era and the scale and extent of contemporary international travel demand the attention of geographers concerned with the study of tourism.

# Origins of international tourism

In Chapter 2 we saw how the development of domestic tourism in Britain followed a clearly defined sequence in which several processes were prominent:

- a spatial development through time of tourist places, from an initial position in which tourism was centred in a limited number of small resorts to an eventual pattern of large-scale development of coastlines and rural hinterlands in which many tourist places may be located;
- a change in motives for travel to resorts from (in the British case, at least) a quest for health to the pursuit of pleasure;
- a process of democratisation of tourism whereby what originates as the exclusive practice of a social elite diffuses down the social ladder to become an important area for mass forms of popular participation.

The development of international tourism also reflects these key processes as exclusive and selective forms of travel have become widely accessible, widely practised and popularised.

Many writers place the origins of modern international tourism in the Grand Tour of the seventeenth and eighteenth centuries. The primary objective of the Grand Tour was to provide young men of wealth and position with the basis of a classical education, by sending them on an extended visit to cultural centres in Europe – in France, Germany, Austria and, especially, Italy. The Renaissance in Europe endowed several nations with a pre-eminence in matters of arts, science and culture, but Italy combined a classical heritage with contemporary ideas and inventions, and its position as an intellectual centre in Europe ensured that for young men of wealth and power, an education could not be considered complete without an extended visit to its main cities. Thus Venice, Padua, Florence and Rome formed a basis to an itinerary that, when extended to include other capitals of culture such as Paris and Vienna, provided the geographical structure for the Grand Tour.

The golden age of the Grand Tour is generally held to be the period between about 1760 and 1790, but references to similar journeys occur much earlier. The Elizabethan courtier Sir Philip Sydney embarked on a tour in 1572, the architect Inigo Jones went to Europe in 1613, the philosopher Thomas Hobbes in 1634 and the poet John Milton in 1638. These tours were probably comparatively short, but by the middle of the eighteenth century a tour might commonly occupy several years.

Although the primary objective remained the completion of a formal education, there were evidently important elements of sightseeing too. Those undertaking the tour would have visited sites of antiquity, art collections, great houses, theatres and concert halls. It also became fashionable to combine travel with the purchase and collection of artefacts: paintings, sculpture, books and manuscripts. Here there are tempting parallels between these early patterns of visiting with their 'souvenir' collecting and later styles of modern tourism.

However, what had started as the preserve of a social elite did not remain so for too long, and by the end of the Napoleonic Wars in Europe in 1815 there was already clear evidence of the emergence of new classes of international traveller, drawn not from the aristocracy but from the bourgeoisie. Because of their more limited budgets, the journey patterns of these new tourists were inevitably shorter and their activities more intensified. Sightseeing became more important than the cultivation of social contacts or the experience of culture. The emergence of new attitudes and ideas at this time also focused the attention of the tourists onto new resources and new tourist places. For example, regions such as the Alps would previously have been characterised as wild and dreadful places, populated by uncivilised peoples and forming major obstacles to travellers *en route* to the important attractions of Italy. However, the romantic movement of the early nineteenth century, and the popularising of the picturesque, transformed public attitudes towards mountain landscapes and quickly promoted new tourist destinations in Switzerland and Alpine zones of France, Italy and Austria. The popularity (and accessibility) of these locations was enhanced still further once organised tours became established. Thomas Cook had created his first European tours in the mid-1850s (mostly to northern France and Germany), but by the mid-1860s had extended his services to include the first package tours to Switzerland and Italy.

As large parts of mainland Europe became populated by an ordinary class of tourist, new areas of exclusive tourism inevitably emerged. Amongst these, the most significant was the French Riviera between Nice and Monte Carlo. Lacking the centres of culture that preoccupied the Grand Tourist, the French Mediterranean coast had escaped the attention of the first tourists, but its attractive coastline and equable climate prompted a process of development that, by the end of the nineteenth century, had established the area as the new pleasure reserve of the European aristocracy. People from the colder climes of northern Europe, in particular, used the French Riviera as a winter retreat, and its visitors numbered most of the crowned heads of Europe and the entourage that always followed people of status.

However, the process of social displacement that is such an apparent aspect of tourism development ensured that ordinary tourists soon followed. The First World War destroyed the social orders that had sustained areas such as the French Riviera as exclusive places, and from the 1920s onwards there is a visible process of social and functional transformation of the French Riviera to a pattern of coastal tourism that was eventually to become widely established along the northern shores of the Mediterranean, drawing both domestic and (especially) international visitors. Initially, the colonisation of the Riviera by influential groups of writers, artists and the new breed of American film stars gave the area a fashionability that was hard to resist. Then, new forms of beach leisure (such as sunbathing – previously a highly unfashionable practice) helped to promote a summer season in an area that had by custom been deemed too oppressive for summer-time visits, whilst new styles of leisure clothing (especially swimwear) reflected a liberalising of attitudes that would soon affect ordinary people. By 1939, the establishment of paid annual holidays in France had brought an influx of lower-class French holidaymakers to the Mediterranean, and the exclusivity of the Riviera had been replaced, in a very short time, by the apparently simple forms of tourism based around sun, sea and sand.

## Post-1945 development of international tourism

The most pronounced developments in the geography of international tourism have, however, been largely confined to the period since the end of the Second World War. During this time there has been unparalleled growth in the number of foreign tourists, a persistent spread in the spatial extent of activity and the associated emergence of new tourist destinations.

## The growth of international tourism

According to the World Tourism Organization (WTO), in 1950 international tourism (as measured in tourist arrivals at foreign borders) involved just 25 million people worldwide – a figure that was no greater

**Figure 3:1**   *Increase in international tourist arrivals, 1950–94*

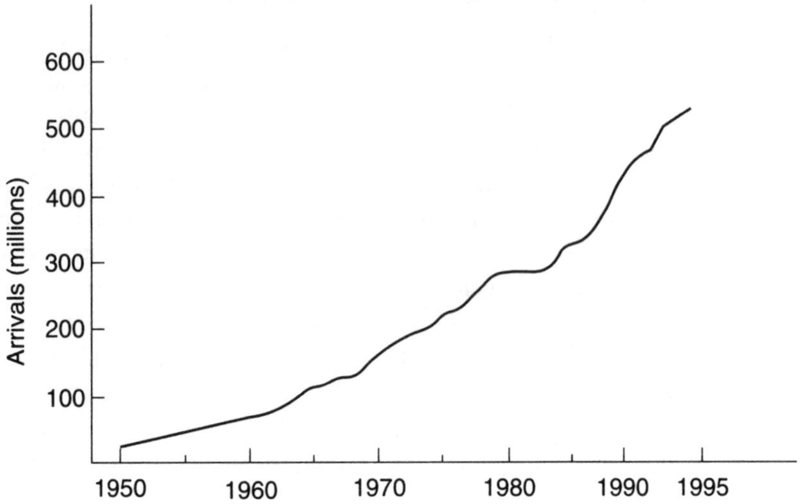

Source: World Tourism Organization (1995).

than the number of domestic holidays taken in a single country, Great Britain, at the same time. From this point, international tourism has risen to an estimated 528 million arrivals in 1994. Figure 3:1 charts the upward trend in detail and suggests two basic features in the pattern of growth.

First, the expansion of international tourism has been almost continuous, reflecting not just the growing popularity of foreign travel but, more importantly, the centrality of tourism within the lifestyles of modern travellers. At a global scale, at least, international tourism appears largely immune to the effects of events that might reasonably be expected to exert an effect. Neither the oil crisis of the mid-1970s, nor the economic recessions of the 1980s, nor the war in the Persian Gulf in the early 1990s appear to have deterred the international tourist to the extent that the upward trends are reversed significantly, although annual rates of increase do show signs of deflection in response to world conditions, especially economic conditions. Thus there occurred a temporary stabilisation of demand in the early 1980s before economic recovery encouraged a further round of growth. But overall, the expansion of international tourism seems irresistible and quite able to withstand pressures of inflation, currency fluctuations, political instability and growing unemployment in most of the countries that generate the principal flows of international tourists.

Second, data show that, unlike many domestic tourism markets which have stabilised or even shown signs of decline (see Figure 2:3, for example), the increase in international travel (when measured in absolute rather than relative terms) is accelerating. Thus in the ten years between 1965 and 1974, the market expanded with an extra 92.8 million arrivals; between 1975 and 1984 it grew by a further 94.7 million; and between 1985 and 1994, by an estimated 200.9 million. With annual growth rates currently running at 4.6 per cent, international tourist arrivals may exceed 690 million by the year 2000.

## The spatial spread of international tourism

Aggregate descriptions, however, conceal a great deal of variation within the basic patterns and these repay closer attention. Historically (and indeed at present), international tourism has been dominated by Western Europe, both as a receiving and as a generating region. This pre-eminence reflects a number of factors including:

- an established tradition in domestic tourism that converts quite readily into international travel;
- a mature and developed pattern of tourism infrastructure, including transportation links, extensive provision of tourist accommodation and organisational frameworks such as travel companies;
- a wealth of tourist attractions including diverse coastal environments, major mountain zones as well as sites of historic and cultural heritage;
- a sizeable industrial population that is both relatively affluent and mobile and thus an active market for international travel;
- a range of climatic zones that favour both summer and winter tourism.

However, as the WTO has noted, although Europeans do possess a higher propensity to travel, the geopolitical structure of the region inflates the level of international travel. In particular, the juxtaposition of relatively small nations creates a large number of international borders that are often routinely crossed by tourists undertaking quite short journeys. In contrast, vacationists in the USA may travel very much further within their home country than the international travellers of Europe, but unless they cross into Mexico or Canada, they fail to register as international travellers.

The extent to which Europe dominates the international tourism market is indicated in Table 3:1, which lists the top fifteen destinations according to visitor arrivals and tourist receipts, alongside the major generators of

**Table 3:1** *International tourism: the major receiving and generating countries, 1991*

| Country (by rank) | Arrivals (millions) | Country (by rank) | Receipts (US$ billions) | Country (by rank) | Expenditure (US$ billions) |
|---|---|---|---|---|---|
| France | 55.7 | USA | 45.6 | USA | 39.4 |
| USA | 42.7 | France | 21.3 | Germany | 31.7 |
| Spain | 35.3 | Italy | 19.7 | Japan | 24.0 |
| Italy | 26.8 | Spain | 19.0 | UK | 18.9 |
| Hungary | 21.9 | Austria | 14.0 | Italy | 13.3 |
| Austria | 19.1 | UK | 12.6 | France | 12.3 |
| UK | 16.7 | Germany | 10.9 | Canada | 10.5 |
| Mexico | 16.6 | Switzerland | 7.1 | Netherlands | 7.9 |
| Germany | 15.6 | Canada | 5.5 | Austria | 7.4 |
| Canada | 15.0 | Hong Kong | 5.1 | Sweden | 6.1 |
| Switzerland | 12.6 | Singapore | 5.0 | Switzerland | 5.7 |
| China | 12.5 | Mexico | 4.4 | Taiwan | 5.7 |
| Portugal | 8.7 | Australia | 4.2 | Belgium | 5.5 |
| Czechoslovakia | 8.2 | Netherlands | 4.1 | Spain | 4.5 |
| Greece | 8.0 | Thailand | 3.9 | Australia | 3.9 |

Source: WTO (cited in Latham, 1994).

international travel. In terms of percentage shares of the world market, in 1993 European countries attracted 60.3 per cent of visitor arrivals and 49.8 per cent of international tourism receipts, whilst one-third of the receipts from tourism at the world level are generated by just ten West European countries.

Within the European area, however, there are marked spatial inequalities, some of which are evident in Table 3:1 and Figure 3:2. Hence Spain, for example, ranks third as a receiving country but only fourteenth as a generator of international tourists, whilst countries such as Germany and the United Kingdom generate more tourists than they receive. Other European countries, especially some in the former Soviet bloc, feature neither as generators nor as receivers of tourists.

The predominant tourist flow in Europe is a north–south movement from the high concentrations of urban-industrial populations in the cooler, northern parts of Europe towards the much warmer areas that fringe the Mediterranean. This helps to establish a Mediterranean 'core' area centred in France, Spain and Italy which dominates the European holiday tourism market and which draws disproportionately upon

Figure 3:2 *Spatial inequalities in tourism in Europe, c. 1990*

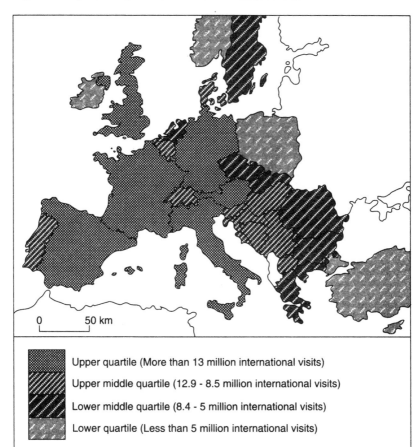

0    50 km

Upper quartile (More than 13 million international visits)

Upper middle quartile (12.9 - 8.5 million international visits)

Lower middle quartile (8.4 - 5 million international visits)

Lower quartile (Less than 5 million international visits)

Germany, the United Kingdom and the Scandinavian countries as sources for visitors. Superimposed upon primary north–south movements are secondary flows to the mountain regions of Europe (both for winter and for summer holidays) and all-season flows between the major European cities for cultural, historic and business tourism. The former trend helps to position both Austria and Switzerland within the top fifteen world destinations, whilst tourism to Britain is strongly dependent upon the latter.

However, other parts of Europe receive much lower levels of international tourism. Even within the Mediterranean there are marked contrasts between east and west, with the Eastern Mediterranean countries receiving only a quarter of the number of visitors to France,

**Figure 3:3** *Changing patterns of air charter tourism from the Republic of Ireland to continental Europe, 1972 and 1991*

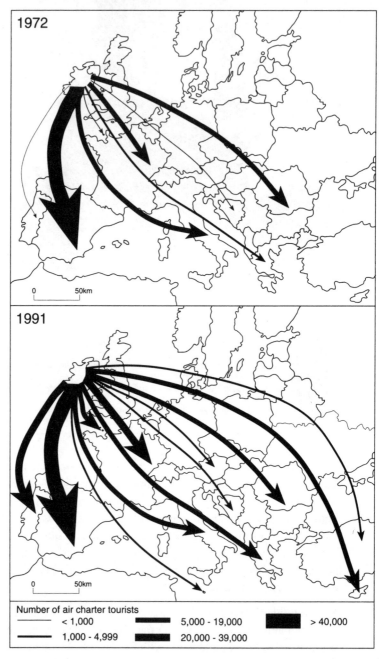

Source: Adapted from Gillmor (1996).

Spain and Italy. Contrasts between tourism in Western Europe and countries of the former Soviet bloc in Eastern Europe are equally pronounced. In 1988 (i.e. immediately prior to the collapse of Soviet influence across Eastern Europe) the combined total of international visitors to East Germany, Czechoslovakia, Poland and Romania was just 17 million, most of whom came from within the East European region itself.

Part of the explanation for differences in levels of tourism in Europe lies in the manner in which the activity has spread. As we have seen, the French Mediterranean coast has a history of tourism which extends back over a hundred years, and the activity appears to have diffused from this region. Thus in the early 1960s, development spread west-wards into Spain and eastwards to the Italian Adriatic coasts. By the early 1970s, the former Yugoslavian coast was an emerging holiday region and tourism to the Greek islands was becoming well established. In the 1980s, package-based coastal tourism reached Turkey. As an example, Figure 3:3 illustrates how patterns of European air charter tourism originating within one state – the Irish Republic – have evolved over a period of twenty years up to 1991. This shows how additional destinations have entered the market whilst countries that formed the initial foci for foreign travel take proportionately smaller shares as the tourist space expands. A process of spatial development of new tourist destinations may also be anticipated in Eastern Europe following the collapse of communism, and there are already signs of rapid growth in tourism to destinations in Hungary and the Czech Republic, notwith-standing the relative weakness of the travel and hospitality industries in these countries at the present.

## New tourist areas

The spatial spread of tourism and the emergence of new tourist areas that may be seen within the European area are also clearly evidenced at the global scale. Although the European share of the world travel market is by far the largest, the trends over the past thirty years show a significant reduction in that share as new, more distant and often more exotic destinations begin to attract the attention of tourists. The nature and extent of these spatial shifts are reflected in Table 3:2 and show clearly how the horizons of international tourism have extended since 1960. Key trends to note are:

- the reduction in the share of the world market of 'established' tourism regions in Europe and the Americas (although the latter grouping is a confusing amalgam of prosperity and growth in tourism in the USA and Canada and almost total underdevelopment of tourism across most of South America).
- the relatively static positions in areas of chronic underdevelopment in the Third World areas of Africa and South Asia (which includes India) and the politically unstable Middle East.
- the dramatic expansion of tourism to East Asia and the Pacific, centred around the thriving tourist economies of Thailand, Singapore, Indonesia, Hong Kong, Japan and Australia.

The scale of the expansion of tourism to these distant locations is emphasised further when these percentage shares of a rapidly developing global market are translated into actual visitor figures. Expressed in this form, tourist arrivals in East Asia and the Pacific have increased from around 690,000 in 1960 to 69 million visitors in 1993, providing a compelling illustration of the extent to which modern tourism has been able to take advantage of shrinking global horizons.

**Table 3:2** *Changing regional distribution of international tourism, 1960–93*

| Region | Percentage share of visitor arrivals | | | | |
| --- | --- | --- | --- | --- | --- |
| | *1960* | *1970* | *1980* | *1990* | *1993* |
| Africa | 1.1 | 1.5 | 2.5 | 3.3 | 3.6 |
| Americas | 24.1 | 23.0 | 21.3 | 20.5 | 20.3 |
| Europe | 72.5 | 70.5 | 66.0 | 62.4 | 60.5 |
| Middle East | 1.0 | 1.4 | 2.1 | 1.6 | 1.6 |
| South Asia | 0.3 | 0.6 | 0.8 | 0.7 | 0.7 |
| East Asia and Pacific | 1.0 | 3.0 | 7.3 | 11.5 | 13.5 |

Source: WTO (cited in Latham, 1994); WTO (1995).

## Factors promoting the growth of international tourism

How have these substantial transformations in the scale and spatial extent of international tourism come about? Explanation for the growth of international tourism needs to consider a wide range of factors, but the main elements may be summarised under the following headings.

# Development of a structured and accessible travel industry

One of the main prerequisites for the growth of international tourism has been the establishment of a mature travel industry, especially since about 1960. This has directly advanced foreign travel in several ways:

- the provision of a wide number of retail outlets (travel agencies) at which foreign travel and holidays may be simply arranged and purchased;
- the development of package tours. These 'commodify' foreign travel by creating inclusive holidays for which travel, accommodation and services at resorts are all taken care of in advance and where the customer buys the holiday as if it were a single product or commodity. Package tours based around air travel – a post-1945 innovation commonly attributed to a Russian *émigré* named Vladimir Raitz, the founder of Horizon Holidays – have been especially significant;
- the provision of good-quality, low-cost or flexible forms of accommodation, for example in budget hotel chains such as the French Formule 1 group or self-catering apartments and villa complexes;
- the provision of local tour and holiday guides who liaise between visitor and host and in so doing often remove or minimise problems that foreign tourists might have with language or custom;
- active promotion of destinations through free brochures and advice services, especially through travel agencies, magazines and newspapers.

As well as sophisticated promotion and selling of foreign holidays, the manner in which the travel industry has been able to encourage or promote improvements in the quality of services offered to tourists at foreign destinations has been an added factor, and it has been noted that through time, the number of places that actively welcome the tourist and reflect that welcome in the development of good-quality accommodation, enhanced transportation services and visitor attractions has also been important.

# Impact of developments in transport and communications

One of the most significant factors enabling the development of international travel has been in the area of transport and communications, especially in the development of commercial air services and, more recently, the acceleration of international rail services and the extension of motorway links.

Air travel is particularly important, and in the field of tourist transport the compression of space and time that the aeroplane has produced has had the most far-reaching consequences for patterns of tourism, ensuring that no part of the globe is now more than 24 hours' flying time from any other part. The advent of jet airliners, and particularly the wide-bodied jets with their increased passenger capacities and extended ranges, halved both journey times and real costs of air travel, and it seems inconceivable that tourism to distant destinations would have grown to the extent that it has if passengers were still being offered the fares, travel times and comfort of the airways of the 1950s. The expansion in international air traffic over the past twenty years has almost exactly matched the expansion in international tourism and, as Table 3:3 shows, recent growth in air travel continues to be healthy and heavily dependent upon tourism to provide the majority of its passengers.

However, the influence of air travel on international tourism patterns is far from consistent. Although estimates suggest that at a global level as many as 35 per cent of international arrivals are by air, in some (major) destination areas air travel holds only a minor share of travel markets. This is indicated in Figure 3:4, which shows the proportions of international visitors who come by air within the WTO areas. Thus, in Europe, although there are strong links between air travel and some sectors of tourism, especially air charter tourism to low-cost Mediterranean destinations, air travel as a whole has a market share that is below the global average and accounts for only just over a quarter of international tourist arrivals. In contrast, tourism to many of the more

**Table 3:3** *Trends in international air transport, 1986–90*

|  | 1986 | 1987 | 1988 | 1989 | 1990 | Percentage change |
|---|---|---|---|---|---|---|
| World tourist arrivals (000s) | 116,321 | 131,833 | 144,514 | 156,779 | 163,549 | 40.5 |
| Passengers carried (000s) | 197,961 | 221,901 | 243,414 | 265,187 | 261,187 | 32.1 |
| Passenger kilometres (millions) | 603,138 | 687,514 | 761,992 | 853,297 | 858,220 | 42.3 |

Source: WTO (cited in Page, 1994).

**Figure 3:4** *Proportion of international tourists arriving by air by region, 1990*

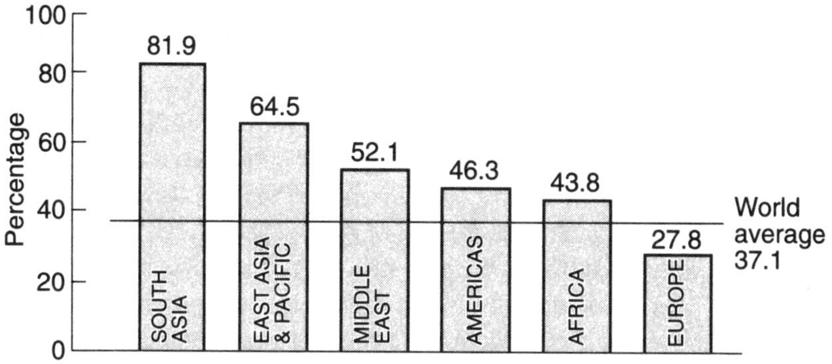

Source: Page (1994).

distant, emerging tourist areas – notably in Asia and the Pacific – is highly dependent upon air travel.

Part of the reason for the secondary significance of air travel in the European area is the convenience of other forms of travel, especially for shorter international journeys. In France, Italy and Spain respectively, 67, 78 and 60 per cent of international arrivals come by car. Travel by road has been aided by developments to international motorways and improvements to the Alpine passes into countries such as Italy, whilst acceleration of rail services, particularly through the Channel Tunnel between Britain, France and the Benelux countries, has added an extra element of competition in sectors such as short-break/city tourism to Paris or Euro-Disney, Brussels and Amsterdam. In the near future it is possible that as much as 20 per cent of short-range air journeys might be lost to these accelerated rail services.

## Economic development and geopolitical stability

International forms of tourism are also dependent upon levels of economic prosperity and geopolitical stability in both generating and receiving areas. Tourism has always been subject to the constraints of cost, and until very recently the expense of foreign travel was a most effective barrier to popular forms of participation. But across large areas of the developed world, general levels of prosperity have risen throughout the post-1945 period, and as levels of disposable income have increased, so foreign travel has become more affordable. The financial accessibility of foreign tourism is also a direct product of the relatively low cost that packaged forms of tourism based around chartered travel will create.

Political stability has an influence too. One of the reasons why international tourism in Europe has developed so strongly since 1945 has been the almost total absence of major political and military conflict in the region since the end of the Second World War. The one significant divide that did arise from that war – the division between a largely communist Eastern Europe and a capitalist West – actually produced a clear demarcation in the geography of tourism, with rapid development in the West and relatively little international travel in the East. As soon as communist control of East European states began to crumble, tourism both to and from these areas followed. The enlargement of the European Union and the gradual erosion of controls on movements between member states will probably extend still further the zones over which international tourism is both encouraged and facilitated.

## The fashionability of international travel

International tourism has developed because it has become fashionable. As we have seen already, the connections between tourism and fashion have often been close, and there seems little doubt that in many contemporary societies, a foreign holiday is a mark of status.

The fashionability of international travel reflects greater public awareness too. Media promotion of travel, through newspapers and magazines, on radio and television, as well as through the travel industry itself, has made people more aware of distant places and, through the construction and dissemination of exotic images of foreign lands, directly promotes the pleasures and experiences that such places can provide. Part of the problem that many domestic resorts now confront is the perception that foreign places will offer an experience that in many ways will be superior – whether it be the enjoyment of a better climate, different landscapes or different places of entertainment, culture, historic or political significance.

## Tourist competence and the ease of foreign travel

Finally, international tourism has expanded because tourists, in general, are more competent at the business of international travel, with changes – both within the industry and in the wider contexts of contemporary economy and society – making foreign travel a much easier process than was once the case:

- post-1945 improvement in educational levels and better training of personnel within the hospitality industries mean that language is less of a barrier;
- travel procedures (customs, airport check-ins, etc.) are rapidly becoming minimised, standardised and familiar;
- computerised reservation systems bring instant access to up-to-date information on availability of flights, rooms or holiday packages and the option of immediate, confirmed bookings;
- credit cards that are valid world-wide simplify financial transactions and purchases whilst minimising the need to carry foreign currencies;
- improved telecommunications make it simpler to keep in touch with home;
- standardised forms of accommodation and other services – in international hotels, restaurant chains and car hire offices – reduce the sense of dislocation that foreign travel might otherwise generate.

The confidence that such familiarity creates is one of the factors promoting the increased tendency to personalised forms of foreign tourism that are largely independent of packaged styles of tourism, and the willingness of tourists to contemplate visiting distant and more exotic locations. As a result, established holiday zones such as the Western Mediterranean coasts are beginning to show signs of decline and an evident need to restructure the types of holidays that they offer to international visitors.

## Variations in patterns of development

However, it is important to appreciate that the factors that have promoted the growth of international tourism vary in their effect through time and across space, producing quite uneven patterns of growth and development. To illustrate this point, three outline case studies of tourism growth – based on Spain, Turkey and Thailand – are now described.

## Spain

Spain is an outstanding example of the impact of post-1945 growth in affordable international tourism and, with an estimated 34 million tourists annually, illustrates a mature destination that may have already reached the final phases of Butler's model (see Figure 2:4). Spain also exemplifies many of the problems that resort areas encounter as they reach their capacities, and the resulting tendency for tourism places to drift down-market, setting in motion a process of spatial displacement of some groups of tourists to new destinations.

**Table 3:4** *Expansion of international tourism to Spain, 1950–90*

| Year | No. of visitors (millions) | Year | No. of visitors (millions) |
|------|------|------|------|
| 1950 | 0.7 | 1985 | 43.2 |
| 1955 | 2.5 | 1986 | 47.3 |
| 1960 | 6.1 | 1987 | 50.5 |
| 1965 | 14.2 | 1988 | 54.1 |
| 1970 | 24.1 | 1989 | 54.0 |
| 1975 | 30.1 | 1990 | 52.0 |
| 1980 | 38.0 | | |

Source: Secretaría General du Turismo (cited in Albert-Pinole, 1993).

Although from the mid-nineteenth century there was a tradition of small-scale local tourism by wealthy Spaniards to resorts such as Málaga, Alicante and Palma de Mallorca, the modern Spanish tourist industry is a visible product of the age of the aeroplane and the international package tour. Spain has benefited from being an early entrant into the field of mass international travel and the period since 1960 has seen rapid and sustained expansion in the numbers of visitors. The data in Table 3:4 are inflated by the inclusion of day excursionists crossing Spanish borders from Portugal and France, as well as migrant workers from North Africa, but the scale of the activity and its spectacular expansion are not in doubt. At its peak of prosperity in the late 1980s, tourism to Spain accounted for 1.2 million jobs (both directly and indirectly), earned US$12.1 billion in foreign currency and generated 11 per cent of Spanish gross domestic product (GDP).

The key factors contributing to the rise of mass forms of tourism to Spain have included:

● the attractive climate;
● the extensive coastline, which includes not just the mainland but also the key island groups of the Canaries and the Balearics;
● the accessibility of Spain to major generating countries in Northern Europe, especially by air;

- the competitive pricing of Spanish tourism products, particularly accommodation, which enabled the extensive development of cheap package holidays to Spanish resorts;
- the distinctive Spanish culture.

Spanish tourism illustrates several features of tourism development, but two are worthy of particular note. The first is the tendency for tourism to concentrate upon particular segments of the market and to be focused into a limited number of locations. Over half of foreign visitors to Spain come from just three countries (France, Germany and the United Kingdom) and they reveal a clear preference for a particular style of low-cost holiday centred on sun, sea and sand. Figure 3:5 shows the distribution of foreign travellers using hotels in 1988 and, with the exception of Madrid, highlights the predominance of the offshore islands and the Mediterranean coastline and the relative insignificance of inland Spain and its northern coastal region. In fact, over 70 per cent of hotel

**Figure 3:5** *Distribution of foreign tourists in Spanish hotels, 1988*

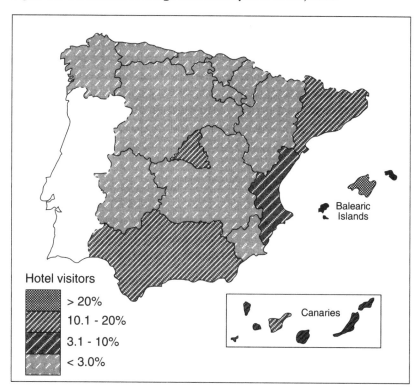

Source: Adapted from Albert-Pinole (1993).

visitors are concentrated into just six regions, and although hotel accommodation houses less than 20 per cent of foreign visitors, the tendency for the major forms of provision (apartments, villa developments, second homes and time-share properties) to focus in the same regions simply exacerbates the spatial unevenness in tourism development.

The second problem is that the rapid pace of development and its spatial concentration have commonly promoted a disorderly pattern of growth. This has undermined the attractiveness of the location, leading to movement down-market. For example, a failure to plan the popular resort of Torremolinos has been one of the elements in its gradual loss of image. Torremolinos, prior to about 1960, was a small fishing village and a resort for a select group of local tourists together with a handful of foreign writers and artists. However, the popularisation of the town as a package tour destination led to rapid and uncontrolled developments which created a formless and untidy built-up area visually polluted by characterless buildings, lacking public open spaces, limited by poor car parking and with an ill-defined and rather inaccessible sea frontage.

Unfortunately, Torremolinos is not an isolated case, and a general incidence in popular Spanish resorts of over-development, commercialisation, crowding of bars, beaches and streets, pollution of sea and beaches as key infrastructure such as sewage treatment has failed to keep pace with expansion, and localised incidence of drunkenness and petty crime have all begun to alter popular perceptions of Spain as a destination. Such problems have become a major source of concern within the Spanish tourism industry.

## Turkey

Turkey is an example of a much newer international destination that may well be benefiting from the displacement of tourists from the established Mediterranean holiday areas such as Spain, as those countries begin to suffer the effects of selective over-development. As noted earlier, Turkey has benefited from the spatial diffusion of tourism from the Western Mediterranean and an eastwards shift in the mass-market 'pleasure periphery' through Yugoslavia and Greece. But the development of Turkish tourism is also a product of active promotion by the industry, the extension of low-cost air travel to more distant destinations, as well as the growing willingness of modern tourists to visit places where the notion of 'difference' is more pronounced.

The recency of Turkey's entry to the tourist industry is mirrored in the fact that prior to 1980, levels of international tourism were minimal. So growth has been highly compressed into a period of a little more than a decade. As a newcomer to international tourism, Turkey has both strengths and weaknesses. The primary strength is the opportunity to learn from the experience of comparable destinations (especially Spain or neighbouring Greece), but the lack of a tradition in tourism, deficiencies in infrastructure and the associated risks that rapid growth will instigate a series of negative impacts upon the economy, society and environment are major areas of weakness. Furthermore, Turkey sits in an ambiguous geographic position: on the edge of Europe (and keenly attempting to become a part of a wider Europe in political and economic terms), yet also facing eastwards in many of its socio-cultural dimensions; for example, its Islamic faith and ethnic composition. Such ambiguity makes the development of a Westernised style of tourism additionally problematic.

Table 3:5 sets out the increase in foreign arrivals between 1982 and 1992. Once again, the data have some deficiencies insofar as figures represent all arrivals, regardless of purpose, but even so, the increase of over 400 per cent in visitor levels in ten years is indicative of the rapidity with which Turkey has entered the international market. Estimates suggest

Table 3:5 *Expansion of international tourism to Turkey, 1982–92*

| Year | No. of visitors (thousands) | Percentage W. Europe | Percentage E. Europe | Percentage other |
|------|------|------|------|------|
| 1982 | 1,392 | 46 | 17 | 37 |
| 1985 | 2,615 | 41 | 21 | 38 |
| 1986 | 2,391 | 52 | 21 | 27 |
| 1987 | 2,856 | 55 | 17 | 28 |
| 1988 | 4,173 | 62 | 14 | 24 |
| 1989 | 4,459 | 60 | 15 | 25 |
| 1990 | 5,389 | 54 | 27 | 19 |
| 1991 | 5,518 | 33 | 53 | 14[a] |
| 1992 | 7,051 | 43 | 45 | 12 |

Source: Economics Intelligence Unit (1993).
[a] Patterns for 1991 affected by proximity of Turkey to conflict in the Persian Gulf.

that in 1992, tourism employed around 250,000 people in Turkey and earned nearly US$4 billion.

Initially, the majority of tourists originated in Western Europe, with as many as 60 per cent of arrivals coming from these countries in the late 1980s. Germany, the UK, the Netherlands and Austria were especially important sources. However, since the collapse of communism and the opening up of travel in Eastern Europe, the balance has altered significantly. In 1982, only 17 per cent of visitors to Turkey came from Eastern Europe, but by 1992 that share had increased to 45 per cent, marginally greater than the level for Western Europe. However, it should be noted that much of the visiting from Eastern Europe is short-term and often unrelated to conventional tourism – for example, travelling salesmen, shoppers and people taking goods for sale in more affluent Turkish markets; hence this activity inflates the overall number of foreign visitors and also misrepresents the true significance of East European tourism in Turkey.

Like Spanish tourism, tourism to Turkey is highly focused in geographic terms. Figure 3:6 shows that nearly 85 per cent of tourist accommodation has developed along the Mediterranean, Aegean and Marmara coasts, leaving the interior and eastern Turkey almost untouched. So although Turkey actually offers a diversity of landscapes, historic and cultural sites, tourism development has tended to imitate and perpetuate the established Mediterranean pattern of sea and beach holidays. To a considerable extent this reflects the levels of foreign control or influence over tourism development that are typically seen in emerging nations and the manner in which the travel industry has chosen to commodify and market the destination. Initial marketing of Turkey to British visitors, for example, was to present Turkey as a low-cost, mass market for seaside tourism. Over 70 per cent of visitors from Western Europe come as air charter tourists on packaged tours of one form or another and stay in either hotels (76 per cent) or holiday villages (15 per cent) in the main coastal provinces.

Unfortunately, the natural temptation for a comparatively undeveloped country to pursue the economic prosperity that international tourism outwardly presents, when compounded by spatial concentrations of activity, encouraged over-rapid and uncontrolled development throughout much of the 1980s. The speed with which arrivals outstripped the supply of both accommodation and supporting

**Figure 3:6** *Distribution of tourist accommodation in Turkey*

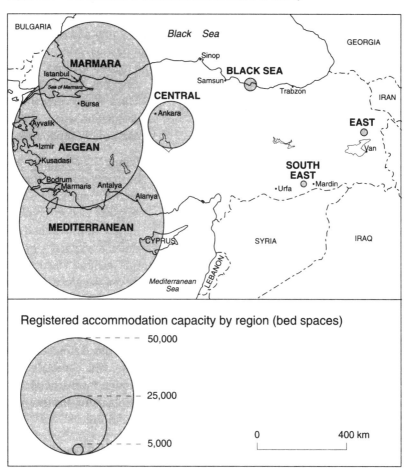

Source: Economics Intelligence Unit (1993).

infrastructure (e.g. water supply, sanitation, visitor services, etc.), quickly endowed the fledgling industry with a reputation for poorly planned, low-quality products. Unsurprisingly, other negative impacts emerged too, including environmental damage through water pollution and poor design of buildings; despoliation of historic sites – both by tourists and local people seeking to cash in by selling souvenirs garnered from sites; and socio-cultural problems, especially where Westernised tourist practices came into contact with orthodox Islamic custom and beliefs.

## Thailand

Thailand exemplifies the globalisation of international travel and the impact of long-haul aviation on geographic patterns of tourism. Located within the WTO East Asia and Pacific region, Thailand represents an emerging tourist destination in a Third World context and a destination where growth has been particularly pronounced.

**Figure 3:7** *Growth of international tourism to Thailand, 1969–94*

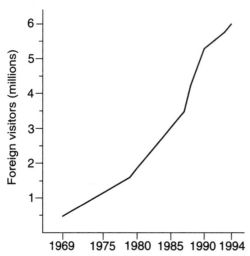

Source: Economics Intelligence Unit (1995); Khunaphante (1992).

Figure 3:7 charts the basic patterns of increase in the foreign visitor market from 1969 – when just 469,000 visitors entered Thailand, mostly from within the South-East Asian area itself – to 1994, when 6.17 million visitors, drawn from across the globe, crossed the Thai borders. The graph also emphasises that the rate of expansion of international tourism to Thailand has tended to increase through time, with especially dramatic growth from around 1987. Explanation for this pattern may be found in:

- the enhanced accessibility by low-cost, long-haul air services from North America and, particularly, Europe;
- the increasing costs in competing destinations (especially the Mediterranean), making Thailand an affordable holiday choice;
- the reduction in the attractiveness of older destinations (for example, Spain) and the counter-attraction of newer, exotic locations;
- the diversity and range of the Thai tourist product, including beach tourism, a distinctive historic–cultural heritage and, less positively, a reputation for unfettered sex tourism and prostitution.

The net effect of this growth has been a significant impact upon the Thai economy. Since 1982, tourism has been Thailand's largest foreign exchange earner, with the industry valued at over US$5.1 billion in 1993. More than 1.5 million Thais are now directly or indirectly employed by tourism.

Table 3:6 *International tourist arrivals in Thailand, 1994*

| Country | No. of visitors (thousands) | Country | No. of visitors (thousands) |
|---|---|---|---|
| Malaysia | 898.8 | USA | 292.3 |
| Japan | 691.7 | United Kingdom | 268.0 |
| Taiwan | 448.2 | China | 257.5 |
| Singapore | 386.9 | France | 219.5 |
| South Korea | 368.4 | Australia | 198.0 |
| Germany | 353.2 | Italy | 130.1 |
| Hong Kong | 310.5 | India | 107.8 |

Source: Economics Intelligence Unit (1995).

The globalised nature of the Thai tourist industry is best reflected in the geographical breadth of the market. Table 3:6 lists the primary markets as defined in 1994 and emphasises the significance of two quite different areas of origin: South-East Asia and Europe. In terms of aggregate numbers of visitors, Asian countries account for 59 per cent of visitors, drawn especially from neighbouring Malaysia, from Japan and noticeably from the then burgeoning economies of Singapore, Hong Kong, South Korea and Taiwan. In contrast, Europe accounted for just 25 per cent of visitors in 1994.

In view of the significance of Western Europe as a source region for international travel, the comparatively small share of the market held by tourists from this region may initially be surprising. However, there are pronounced differences in the lengths of stay in Thailand, with the average Asian visitor stopping for less than 5 days whilst the typical European tourist stays for more than 11 days. So when figures are recast to express market shares in terms of visitor-days, the importance of the European tourist is greatly increased. (Asian visitors account for 42 per cent of visitor-days to Thailand; European tourists account for 39 per cent.)

Thailand demonstrates a common feature of the spatial incidence of international tourism whereby different groups of tourists occupy different places. As in both Spain and Turkey, tourism to Thailand is concentrated into particular localities, and extensive tracts of land, especially in northern Thailand, are relatively untouched by either domestic or international tourism. Bangkok, as the capital, the main port of entry for airline visitors, the location of many of the

international-standard hotels and the site of a number of major
attractions, is an important destination for European and Japanese
visitors, and there are major Westernised beach-resort zones in south-
central Thailand around Phuket and the island of Ko Samu. However,
tourism to provinces in the far south of Thailand shows a quite different
pattern of tourism, with a market dominated by young Malays and
Singaporeans making short visits across the border, attracted by the
cheap shopping and the nightlife. In this locality, European tourists are
almost totally absent.

A spatial unevenness in the development of the industry is not, however,
the only problem with which tourism in Thailand is faced, and longer-
term growth may well be dependent upon solutions to the following
issues:

● the inevitable deficiencies in basic infrastructure in a less developed
  country, especially away from Bangkok and the major tourist places;
● the relative immaturity of the industry, which is reflected, for example,
  in the dependence upon hotel accommodation, much of which is
  foreign-owned. (Ninety-seven per cent of foreign visitors to Thailand
  stay in hotels.)
● the quality of many Thai tourist products. Shopping is a major tourist
  pastime in Thailand but the activity is beset with problems of fraud
  (especially in jewellery and souvenirs) and the retailing of smuggled
  goods;
● a lack of control over physical development, including over-
  development of some sites and a general absence of land-use zoning,
  which has begun to create negative environmental impacts, serious
  traffic congestion and associated pollution in major cities;
● some deterioration in the image of Thailand as a destination,
  especially associated with the country's growing reputation as a centre
  for international sex tourism and prostitution.

The experience of each of these destinations shows just how rapidly
international tourism has tended to develop, and equally, how quickly
potentially damaging changes can be instigated. It also highlights the
practical difficulty of regulating such a diffuse industry in which
unchecked and spontaneous forms of development occur all too easily
and where the positive benefits that tourism can bring are often
outweighed by negative impacts. In all three cases, problems within the
tourism industry have necessitated varying levels of intervention by
governments and governmental agencies in the development process, in

an attempt to provide appropriate forms of control and direction to tourism, and it is to the broader themes of development and associated impacts that we now turn.

# Summary

The theme of this chapter is the spatial expansion in tourism as evidenced in the development of international travel. Although rooted in the history of the Grand Tour, international tourism is shown to be primarily a product of post-1945 patterns of leisure where growth has been aided by a range of factors. These include:

- the development of a structured travel industry;
- the impact of technological innovation in transportation and communications that has altered the speed and costs of travel to the benefit of tourist;
- economic and political stability;
- the fashionability and ease of foreign tourism.

Whilst Europe still dominates international tourism markets, the recent development of new tourist areas in so-called 'long-haul' destinations in East Asia and the Pacific suggest that new tourism geographies are already emerging.

# Discussion questions

1  What do you understand by the phrase 'the democratisation of travel' and how may this concept be used to explain developing patterns of international tourism after 1850?
2  To what extent are spatial patterns in international tourism a reflection of changing tastes and fashions?
3  How have organisational developments in the travel industry assisted the growth of international tourism?
4  What does the experience of destinations such as Spain, Turkey and Thailand suggest are likely to be the primary development problems associated with international tourism?

# Further reading

Despite its age, one of the best accounts of the development of international travel remains:

Turner, L. and Ash, J. (1975) *The Golden Hordes: International Tourism and the Pleasure Periphery*, London: Constable.

whilst an excellent recent discussion of development up to 1940 is provided by:
Towner, J. (1996) *An Historical Geography of Recreation and Tourism in the Western World, 1540–1940*, Chichester: John Wiley.

For a wide-ranging discussion of tourism developments in Europe, see:
Pompl, W. and Lavery, P. (eds) (1993) *Tourism in Europe: Structures and Developments*, Wallingford: CAB International.

whilst texts with useful European case studies and examples are included in:
Davidson, R. (1992) *Tourism in Europe*, London: Pitman.
Williams, A.M. and Shaw, G. (eds) (1988) *Tourism and Economic Development: Western European Experiences*, London: Belhaven.

For a collection of essays on tourism development in the Third World see:
Harrison, D. (ed.) (1994) *Tourism and the Less Developed Countries*, London: Belhaven/John Wiley.

Impacts of transport upon international travel are discussed by:
Page, S. (1994) *Transport for Tourism*, London: Routledge.

For a selection of case studies of specific destinations see, *inter alia*:
Albert-Pinole, I. (1993) 'Tourism in Spain'. In Pompl, W. and Lavery, P. (eds) (1993): 242–261.
Cooper, C.P. and Ozdil, I. (1992) 'From mass to "responsible" tourism: the Turkish experience', *Tourism Management*, Vol. 13 No. 4: 377–386.
Economics Intelligence Unit (1993) 'Turkey', *International Tourism Report* No. 3: 77–97.
—— (1995) 'Thailand', *International Tourism Report* No. 3: 67–81.
Gomez, M.J.M. (1995) 'New tourism trends and the future of Mediterranean Europe', *Tijdschrift voor Economische en Sociale Geografie*, Vol. 86 No. 1: 21–31.
Pollard, J. and Rodriguez, R.D. (1993) 'Tourism and Torremolinos: recession or reaction to environment?' *Tourism Management*, Vol. 14 No. 4: 247–258.
Valenzuela, M. (1988) 'Spain: the phenomenon of mass tourism'. In Williams, A. and Shaw, G. (eds) (1988): 40–60.

# 4 ▶ Costs or benefits? The physical and economic development of tourism

Amongst the many impacts that tourism may exert upon host areas, the processes of physical and economic development are perhaps the most conspicuous. These effects may be evident in the physical development of tourism infrastructure (accommodation, retailing, entertainment, attractions, transportation services, etc.); the associated creation of employment within the tourism industry; and, less visibly, a range of potential impacts upon GDP, balances of trade and the capacities of national or regional economies to attract inward investment. For developing regions in particular, the apparent capacity for tourism to create considerable wealth from resources that are often naturally and freely available has proven understandably attractive, but the risks associated with over-development and dependence upon an activity that can be characteristically unstable are negative dimensions that should not be overlooked. There are benefits, but there are also costs attached to the physical and economic development of tourism.

For the student of tourism geography, however, 'development' itself can be a problematic concept. This is due primarily to the diversity of ways in which the term has been applied – describing both a *process* of change and a *state* (or a stage) of development. Thus, for example, Butler's model of the resort cycle (see Chapter 2) essentially defines successive stages of development but does not, in itself, articulate the details of process. Further, not only is there a basic distinction to be made between state and process, but the nature of the process has been subject to a variety of interpretations, including, *inter alia*:

- development as a process of economic growth, as defined in increased commodity output, creation of wealth and a raising of levels of employment;
- development as a process of socio-economic transformation in which economic growth triggers wider processes of change that alter relationships between locations (particularly between developed and underdeveloped places) and between socio-economic groups – thereby creating fundamental shifts in patterns of production and consumption;
- development as a process of spatial reorganisation of people and areas of production. This may be viewed as a visible product of socio-economic transformations and is a common adjunct of tourism development, with its propensity to focus attention upon resources and resource areas that may previously have been idle or little used.

Within geography, development studies have traditionally tended to explore the particular problems of less developed states and their relationships with the developed world. Part of this tradition has also transferred to the geographic study of tourism, but it is important to note that tourism development processes are also highly significant within states that would already be described as 'developed'. This is a natural reflection of the recency with which mass forms of tourism have emerged, and whilst some parts of this chapter will examine tourism in the context of less developed nations, most of the discussion is concerned with physical development and economic impacts within the settings of developed nations.

The chapter is cast into two distinct, but related, sections. Initially, discussion is centred upon the factors that shape and regulate the physical development of tourism and the contrasting spatial forms that may result. Then the wider relationships between tourism and economic development (including the capacity to create employment) are introduced and explored.

# Patterns of physical development of tourism

## Prerequisites for growth

The development of tourism in any given location requires that several key elements come together to produce the right conditions. These may be summarised under three headings: resources and attractions; infrastructure; and investment, labour and promotion.

## Resources and attractions

Tourism is a resource industry, dependent for its basic appeal upon nature's endowment and society's heritage. The natural appeal of a locality may rest upon one (or more) of its physical attributes: the climate, landforms, landscapes, flora or fauna; whilst socio-cultural heritage may draw tourists seeking to enjoy centres of learning or entertainment, to visit places of interest or historic significance or to view buildings or ruins of buildings. Socio-cultural attractions may also extend to the perusal of artefacts or works of art; the experience of customs, rituals or performing arts; enjoyment of foreign cuisine; or festivals and spectacles.

In addition to the natural and social endowments of an area, the industry will typically seek to develop the resource and attractions base to tourism through the construction of specific, often artificial, tourist attractions. Examples might include tourist shops, places of entertainment and amusement, theme parks, swimming pools and leisure complexes.

## Infrastructure

Tourism development requires infrastructure, primarily in the form of accommodation, transportation services and public utilities. Tourism, by definition, is centred upon travel and on staying away from home, hence the provision of both transportation and accommodation will be integral elements within development programmes. Transportational developments need to take account of the needs for external linkages (ports, airports, international rail terminals, etc.) to allow tourists to gain access to their destinations, as well as provision that allows for circulation within the destination area (local roads, vehicle hire services, etc.). Accommodation developments may reflect particular market segments at which the destination is being targeted (for example, luxury hotels for discerning international travellers), but otherwise must cater for the diversity of tourism demands by providing not just serviced accommodation in the form of hotels, but also cheaper or more flexible forms of accommodation: in apartment blocks, villa developments, time-shares or caravan and camping sites. The expectations of quality that many tourists carry with them also have implications for provision of public utilities; water supply, sanitation and electricity are essential underpinnings to most forms of modern tourist development.

## Investment, labour and promotion

For a tourism area to develop, there is a need for sources of capital investment, labour and appropriate structures for marketing and promoting the destination to be established. Whilst some of the basic attractions to tourists (especially the natural phenomena) may in a sense be 'free', infrastructural developments and the formation of artificial attractions require investment, and the operation of the industry at the destination requires pools of labour with appropriate training and experience. In most developmental contexts, such needs are met by combinations of private and public investment, with governments typically playing a greater role in the promoting of destinations, in infrastructural improvements involving transport and public utilities, and, in some cases, in employment training. In contrast, private finance is more prominent in the development of tourist accommodation and attractions. However, the balance between public and private finance (and between indigenous and foreign investment) will vary considerably from place to place, depending upon local economic and political conditions.

## Factors shaping physical development of tourism

However, although the prerequisites that shape tourism development may be defined without too much difficulty, different *processes* of development can and do occur, and these will be associated with contrasting spatial patterns and forms.

Figure 4:1 attempts an outline summary of what is actually a most complex set of processes and suggests that the contrasting spatial forms of tourism development (the 'development outcomes') may be viewed as a product of interplay between 'factors of influence'. Five primary factors are proposed:

- physical constraints;
- the nature of tourist resources and attractions;

**Figure 4:1  Factors affecting patterns of tourism development**

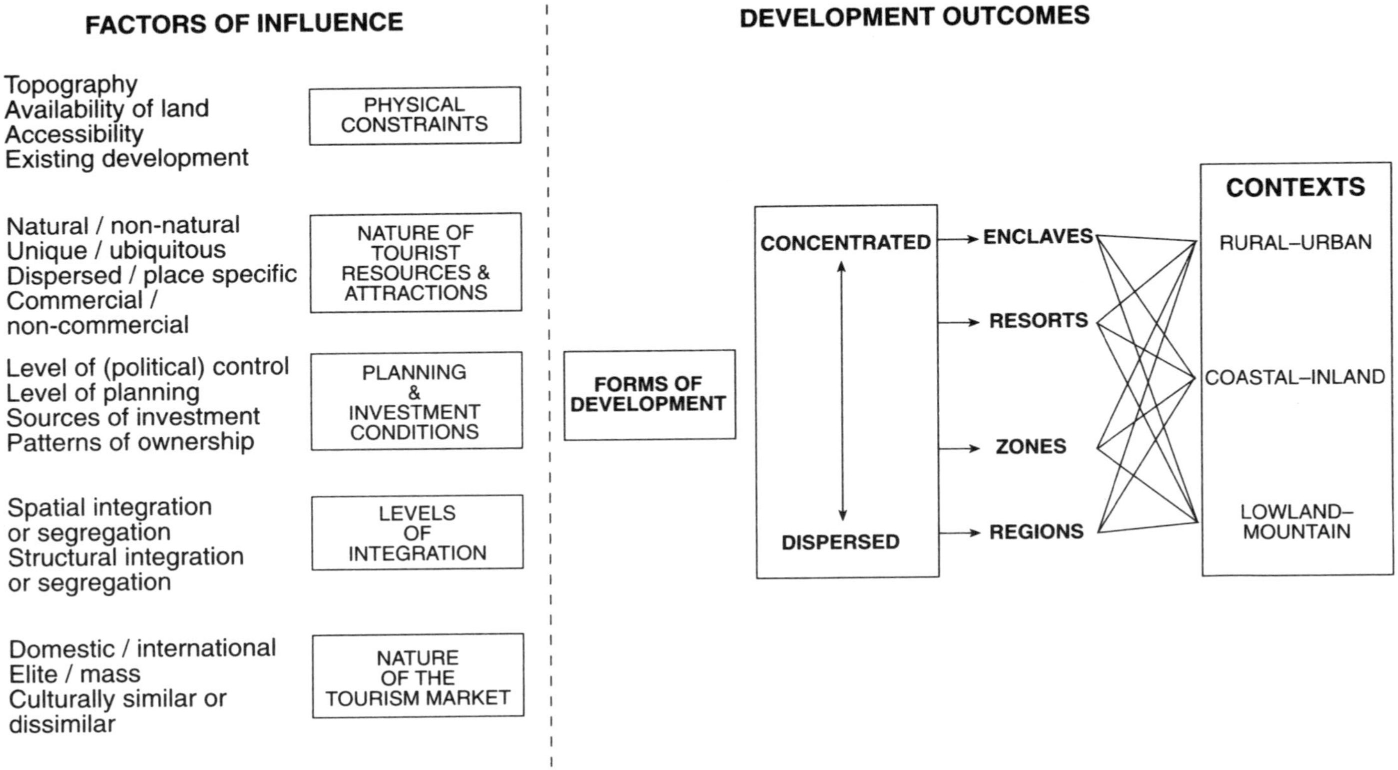

FACTORS OF INFLUENCE

DEVELOPMENT OUTCOMES

CONTEXTS

RURAL–URBAN

COASTAL–INLAND

LOWLAND–MOUNTAIN

ENCLAVES

RESORTS

ZONES

REGIONS

CONCENTRATED

DISPERSED

FORMS OF DEVELOPMENT

PHYSICAL CONSTRAINTS

Topography
Availability of land
Accessibility
Existing development

NATURE OF TOURIST RESOURCES & ATTRACTIONS

Natural / non-natural
Unique / ubiquitous
Dispersed / place specific
Commercial / non-commercial

PLANNING & INVESTMENT CONDITIONS

Level of (political) control
Level of planning
Sources of investment
Patterns of ownership

LEVELS OF INTEGRATION

Spatial integration or segregation
Structural integration or segregation

NATURE OF THE TOURISM MARKET

Domestic / international
Elite / mass
Culturally similar or dissimilar

- planning and investment conditions;
- levels of integration;
- the nature of the tourism market.

The spatial forms that result from this interplay are conceived as falling broadly into four categories: enclaves, resorts, zones and regions. In spatial terms, these different forms are associated with varying levels of concentration or dispersal and may also be located into one of several geographic 'contexts' that are here expressed as simple continua: urban/rural; coastal/inland; lowland/mountain, etc.

Each of the primary factors are themselves made up of more specific influences that may be briefly elaborated. First there are sets of physical constraints that will have a direct bearing upon forms of development. Topography, for example, can influence the availability of suitable sites for construction, levels of access and the ease with which key utilities (water, power, sewage disposal, etc.) may be installed or extended from existing settlements and their infrastructure. 'Difficult' environments include rugged coastlines or mountain zones, both of which tend to fragment and disperse development in a way that is generally untrue of (say) a flat, open coastline which enjoys ease of access.

Second, development patterns will reflect characteristics of the resources and attractions around which tourism is based, and affect especially the extent to which tourism becomes dispersed or concentrated. In particular, unique or place-specific attractions, whether natural or non-natural, tend to focus development around the site(s) in question, whereas more ubiquitous or spatially extensive resources (for example, an accessible coastline or good-quality rural landscapes) may have a dispersing effect. Thus, rural tourism – in which sightseeing is an important pastime – is often characterised by a diffuse pattern of development at a multiplicity of relatively small-scale sites, with activity frequently being absorbed within existing facilities through farm tourism or second homes (where these are conversions of existing properties).

Although, historically, many forms of tourism development were spontaneous and only loosely controlled, the value of tourism as a tool for regional and national development has tended to mean that the modern industry is far more closely regulated. Local planning and investment conditions will therefore be a third primary influence upon forms of development, and, as Figure 4:1 suggests, important factors include political attitudes towards tourism and the levels of political control (including the extent to which effective land planning procedures are in place); the extent to which investment is local or external to the region; and the levels of corporate interest in tourism and the associated patterns of ownership.

Planning and investment conditions are closely allied with a fourth key factor, the level and nature of integration. Discussions of 'integration' of tourism development tend to use the term in two senses. At one level, concerns have focused upon the extent to which tourism development is integrated in a spatial sense with existing, non-tourist forms of development – in other words, is tourism inter-mixed with other functions and land uses, or is it spatially segregated? Alternately, integration may refer to whether or not a development is integrated in a structural sense. A structurally integrated development will bring together all the key elements – accommodation, transportation, retailing, entertainment and utilities – within a single, comprehensive development. This form contrasts with what are sometimes termed 'catalytic' patterns of development in which a small number of lead projects, which are often externally financed and controlled, stimulate subsequent rounds of indigenous development as local entrepreneurs are drawn into an expanding tourism industry.

Finally, it is suggested that patterns of development will be influenced by the nature of the tourism market. They will vary according to whether development is targeted at a domestic or an international clientele, but more significant distinctions will normally exist between elite and mass forms of tourism, whilst levels of cultural similarity/dissimilarity between host and visitor may also be reflected in the manner in which development is organised.

## Contrasting forms of tourism development

We can exemplify how these different elements interact to produce varying forms of tourism development by examining the three most common development 'outcomes': tourist enclaves, resorts and zones.

### Tourist enclaves

Enclaves represent the most highly concentrated form of tourism development and reflect most clearly the influence of:

- the constraints posed by limitations in infrastructure within a locality;
- investment patterns in which there are relatively few entrepreneurs developing provision for tourists and where funding is likely to be external in origin;
- a market which is focused upon a particular segment – usually elite groups – and where the tourist activity is often concentrated upon a particular resource – usually beach resorts.

Enclave developments, in their purest form, are entirely enclosed and self-contained areas, not just as physical entities, but as social and economic entities too. They will display several features:

- physical separation (and isolation) from existing communities and developments;
- a minimising of economic and other structural linkages between the enclave and the resident community;
- a dependence upon foreign tourists which is reflected in pricing structures that reinforce the exclusivity of the enclave;
- pronounced lifestyle contrasts between the enclave and its surroundings.

Enclave developments are often a reflection of immaturity (or a pioneering stage) within a local tourism industry that has yet to evolve to the point where it can support a wider base of provision. In this sense, Regency Brighton, for example, represented a leisure enclave – a socially exclusive space that through real and symbolic boundaries was accessible only to a favoured few. However, in modern tourism, enclaves are most commonly found in developing nations (see Box 4:1), although this is not exclusively the case. The recent development in temperate parts of Europe (e.g. Britain, Belgium and the Netherlands) of high-quality indoor holiday villages with integral and comprehensive facilities set in artificially regulated 'exotic' environments marks a reworking of the enclave idea that is very much the product of a developed rather than a developing economy. For the moment, however, these particular forms of development, of which the Dutch firm Center Parcs (now British owned) is perhaps the leading exponent, are exceptional.

For tourism in emerging nations, enclave developments offer several distinct advantages. First, the concentration of investment into small numbers of contained projects represents a pragmatic response to the problems of how to begin to provide the high-quality facilities that modern travellers expect and how to form and reinforce a distinct and marketable product. Second, the tendency for enclaves to be partially or

## Box 4:1

### Tourism enclaves in the Dominican Republic

The Dominican Republic lies within the Caribbean, occupying the eastern part of the island of Hispaniola, which it shares with Haiti. The land area is some 48,733 km², and the population is 7.1 million.

In relation to some Caribbean states (e.g. Barbados or Jamaica), the Dominican Republic is a relatively late entrant to the world of international tourism. Concerted government efforts to promote tourism as a means of diversifying the economy and raising the extremely low standards of living of the majority of Dominicans emerged only in the 1970s. At first, development was largely funded by domestic investors, but as the tourism industry has begun to expand, higher levels of foreign investment have been noted, especially from the USA.

The tourism development strategy has been to focus investment into six designated tourism zones within which enclave resorts are the preferred development model (Figure 4:2). This is a conscious decision aimed at rationalising the use of scarce funds on infrastructural improvements that have included a new airport, road developments and sewage treatment schemes.

The Luperon Beach Resort, located at the western end of the Puerto Plata tourist zone, is an example of one of these enclaves. The resort is a 160-room

**Figure 4:2  Tourism development areas in the Dominican Republic**

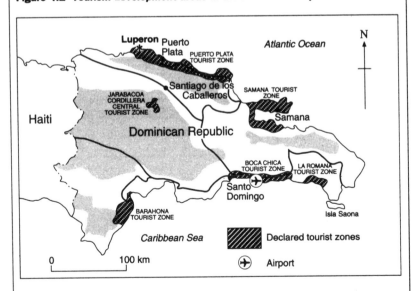

Source: Adapted from Freitag (1994).

development of international-standard accommodation with integral bars, restaurants and shops retailing drinks, tobacco and souvenirs. At the height of the season, more than 100 local people are employed in the resort, servicing the needs of, in the main, package tourists – albeit of some affluence.

However, although the resort enjoys high occupancy rates and is generally profitable, the development has been associated with a range of negative impacts that highlight the potential risks for host communities of enclave development. In particular, the packaging of the resort as an 'all-inclusive' destination (in order to maximise profits) has minimised levels of spin-off trade entering the local economy. Many goods and services are purchased in bulk from major Dominican centres or are imported directly from offshore suppliers, whilst tourist demands that are met through local sources often create shortages for the community. Employment, though important, is generally seasonal, and few local workers secure positions of responsibility within the resort. Such patterns are generally characteristic of enclave tourism in less developed countries.

Source: Freitag (1994).

often entirely financed and owned by offshore companies is seen as a means of attracting inward investment to the developing economy and creating service employment for local people. Third, and less obvious, is the fact that enclaves may be favoured by local governments that are anxious to contain or limit potentially adverse social, cultural or political effects emanating from contact between visitors and host populations.

However, set against these potential benefits are several serious weaknesses, including increased economic dependence on foreign corporate institutions and investors; high levels of 'leakage' from the economy – especially in the form of profit paid to foreign owners or investors; limited levels of dependence upon local supplies of goods and services; and, sometimes, a seasonality in the employment of labour. These problems are explored more fully in the second half of this chapter.

## Resorts

The most familiar form of tourism development is the resort. Resorts may occur in a number of contexts. The seaside resort is the most commonplace, but resorts may also develop around inland health spas (e.g. Harrogate, England), in mountain regions (e.g. La Grande Plagne, France) and even in deserts (e.g. Palm Springs, California, and Las Vegas, Nevada). Resort developments are perhaps most strongly influenced by the nature of the resources that form the basis of their attraction, and therefore a concentrated form of development tends to occur, centred around key resources; but at a detailed level, resorts will also illustrate the effect of accessibility and availability of land, levels of planning and control, sources of investment and varying levels of integration.

As it is the most established form of development, we will concentrate upon the seaside resort. The historic evolution of seaside resorts has already been traced in Chapter 2, and the processes described there have produced a form of resort development that may be considered 'traditional' and which is widely encountered in countries such as Britain. Such resorts reveal the attraction of the sea as a resource, whilst their complex land patterns point to processes of incremental growth, much of it spontaneous and unplanned, and often in the form of small-scale, local investment.

In general, the importance of the sea within these resorts has been responsible for a common pattern of linear development along the sea frontage itself, with pronounced 'gradients' of decline in land values and associated changes in land uses with increasing distance from the front. A secondary gradient of change is also evident *along* the front as one moves from prime locations at the core of the resort towards the periphery. The natural tendency for certain resort functions – accommodation, tourist retailing and entertainment – to group together for commercial reasons then produces quite well-defined spatial zonations within the traditional resort, albeit integrated and interspersed with other non-resort activities; for example, local industry or residents' housing. Zoning may also lead to the formation of a distinctive 'recreational business district' (RBD) that is partially or wholly separate from the normal 'central business district' (CBD) that we may expect to find in any urban place. Furthermore, within tourist zones, competition for prime sites will tend to separate larger enterprises from smaller ones, whilst the particular needs of some sectors (for example, the need for bed and breakfast houses to attract passing trade) will produce particular locational tendencies within sectors. In the case of bed and breakfast houses, the attraction of positions on main routeways is an observable pattern. For different reasons, low-cost functions that require large areas of land (for example, holiday villages and caravan parks) gravitate to the edge of the resort where cheap land is most likely to be available. Figure 4:3 provides a diagrammatic summary of these ideas in the form of a simple descriptive model.

**Figure 4:3** *Model of a conventional seaside resort*

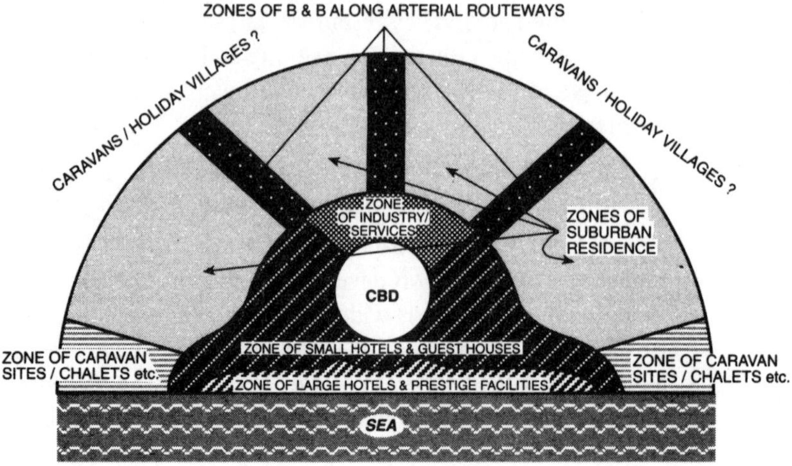

However, as tourism has developed, different forms of beach resort have emerged that do not match the 'traditional' model since development conditions may be different and the extended chronology of development that has shaped the traditional resort is absent. Figure 4:4 summarises a model of beach resort development based upon empirical observation of modern resort formation in the Asia–Pacific region.

This more complex description repays closer attention, but distinctive features that are worth emphasising include:

- the role of second-home development in the early phases of resort development;
- the tendency initially towards strip development along the sea frontage which is reinforced by the first phases of hotel development;
- the processes of displacement of residential properties from the frontage as the tourism industry becomes established;
- the emergence of secondary developments of hotels at inland locations, once the front has become fully developed;
- the eventual separation of a CBD from an RBD in the mature stages of the resort's formation.

Although developed around observation of resort formation in Malaysia, Thailand and Australia, this model can also be applied to some forms of resort in other tourism regions, including the Mediterranean. Box 4:2 presents an outline example of one of the beach resorts studied in the formation of this model.

**Figure 4:4** *Smith's model of beach resort formation*

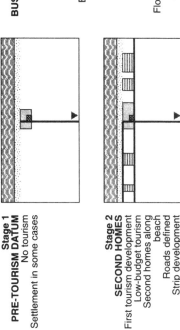

**Stage 1**
**PRE-TOURISM DATUM**
No tourism
Settlement in some cases

**Stage 2**
**SECOND HOMES**
First tourism development
Low-budget tourism
Second homes along beach
Roads defined
Strip development

**Stage 3**
**FIRST HOTEL**
Visitor access improved
First hotel opens
Ad hoc development
High-budget visitors
Jobs in tourism

**Stage 4**
**RESORT ESTABLISHED**
More hotels
Strip development intensified
Some houses displaced
Residential expansion
Hotel jobs dominate

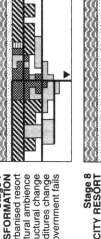
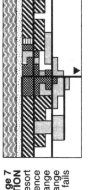
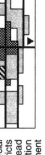

**Stage 5**
**BUSINESS AREA ESTABLISHED**
More accommodation
Visitor type broadens
Non-hotel business growth
Tourism dominates
Large immigrant workforce
Cultural disruption
Beach congestion and pollution
Ambience deteriorates

**Stage 6**
**INLAND HOTELS**
Hotels away from beach
Rapid residential growth
Business district consolidates
Flood & erosion damage potential
Tourism culture dominates
Traditional patterns obliterated
Entrepreneurs drive development
Government master plan

**Stage 7**
**TRANSFORMATION**
Urbanised resort
Rehabilitation of natural ambience
Accommodation structural change
Visitors and expenditures change
Resort government fails

**Stage 8**
**CITY RESORT**
Fully urbanised
Alternative circulation
Distinct recreational and commercial business districts
Lateral resort spread
Serious pollution
Political power to higher government

Sea    Beach    Businesses    Hotel    Second homes    Residential    ▬ Road    ▼ Prime access

Source: Reprinted from *Landscape and Urban Planning*, Vol. 21, Smith, R.A. 'Beach resorts: a model of development evolution', pp. 189–210. Copyright (1991), with kind permission from Elsevier Science Ltd, The Boulevard, Langford Lane, Kidlington, OX5 1GB.

## Box 4:2

### *Development of a modern beach resort: Pattaya, Thailand*

In the 1940s, Pattaya (which lies 140 km south-east of Bangkok) was a relatively inaccessible fishing community which contained a handful of second homes established by wealthy Thais. These second homes formed the basis of the development of the resort, and as more were added they formed a distinct zone that has survived within the contemporary resort (see Figure 4:5). Improved road access to Bangkok in the early 1960s, which coincided with the establishment of US military bases in the region, created new demands that led to the construction of the first hotels on the beach frontage from 1964. These provided the catalyst to a subsequent expansion of hotel-based tourism catering for both the domestic

**Figure 4:5**   *Pattaya, Thailand: resort location and structure*

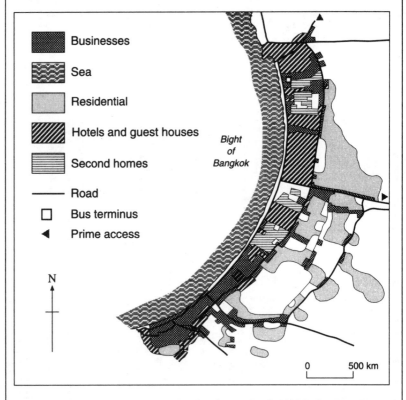

Source: Reprinted from *Landscape and Urban Planning*, Vol. 21, Smith, R.A. 'Beach resorts: a model of development evolution', pp. 189–210. Copyright (1991), with kind permission from Elsevier Science Ltd, The Boulevard, Langford Lane, Kidlington, OX5 1GB.

market and visitors from Europe and Australia that, since the mid-1980s, has seen significant growth with prominent zones of hotels and guest houses, especially along the northern shores. Meantime, as the model hypothesises, business functions have developed strongly, whilst the expansion in the resident population base – much of it drawn by the growth of the tourism industry – has established areas of residential use inland from the coast.

Source: Smith (1991)

## Tourism zones

In mature tourism destinations, the scale and extent of development will often proceed to the point at which extended zones of tourism emerge. These typically will be formed by combinations of resorts, enclaves and other types of development (for example, villa complexes, holiday villages, caravan sites, attractions, golf courses, etc.) to provide a landscape that is infused with tourism. In contrast to the other forms discussed above, however, the emphasis in zonal development is upon dispersal rather than concentration, although there may still be concentrations of activity *within* the zone.

The precise form that such zones may take is variable, reflecting key factors of topography, access, availability of land for development, and planning and investment conditions. However, one of the most characteristic patterns of zonal development is a linear growth along accessible and attractive coastlines. In some instances the topography encourages such growth by creating only narrow coastal strips that are suitable for development, but the attraction of the seashore also tends to encourage linear forms, irrespective of physical constraints. This tendency may then be further reinforced by, for example, construction of coastal roads that link the different elements together. In conditions where local planning control is poor, development will tend to be spontaneous and produce anarchic patterns in which negative impacts will often be pronounced. Box 4:3 provides an example of linear-zonal development from the Spanish Mediterranean coast.

## Tourism and economic development

The physical development of tourism is, of course, linked with a range of environmental and social impacts (see Chapters 5 and 7), but the closest ties are arguably economic in character. Tourism may:

## Box 4:3

### *Linear zonal development: the Costa del Sol, Spain*

The Spanish Costa del Sol provides an excellent example of how linear forms of tourism development may be formed (see Figure 4:6). Prior to the popularisation of cheap package tours to Spain, this coastline was relatively unvisited by tourists, but after 1960 a rapid and largely unchecked expansion of facilities created continuous (or semi-continuous) forms of urban development along the coastal strip.

Initially, expansion was focused upon the resorts of Torremolinos, Marbella, Fuengirola and the main point of entry to the region by air, Málaga. However, as visitor numbers increased, there was a diffusion of tourism from major resorts to secondary centres, which in turn developed rapidly, in some cases to the point at which adjacent places fused to form new urbanised zones, for example Torremolinos and Benalmadena. This process was assisted by the upgrading of the main coast road (N340), which has permitted the spread of tourism westward from the original sources of development near Málaga to the more westerly parts of the coast around Estepona.

However, development of this coastline is not merely centred within the resorts, and the linear qualities have been further emphasised by processes of in-filling of land around and between established centres. Some of this has been in the form of camp sites, some land has been given to the development of golf courses, but most conspicuous has been the growth of what the Spanish term *urbanizaciones*. These are villa developments and include second and/or retirement homes belonging to both Spanish and foreign owners, as well as houses for rent by tourists. There are now more than 600 *urbanizaciones* spread along the coast and the lower slopes of the coastal hills in disorganised and fragmented patterns, and it is these elements, as much as any, that have cemented the linear style of tourism development that is such a distinctive feature of this coastline.

Source: Pearce (1987); Patronato Provincial de Turismo de la Costa del Sol (1988).

- aid economic development through the generation of foreign exchange earnings;
- exert beneficial effects upon balance of payments accounts;
- create substantial volumes of employment;
- assist in the redistribution of wealth from richer to poorer regions;
- promote and finance infrastructural improvements;
- diversify economies and create new patterns of economic linkage.

**Figure 4:6  Tourism development on the Spanish Costa del Sol**

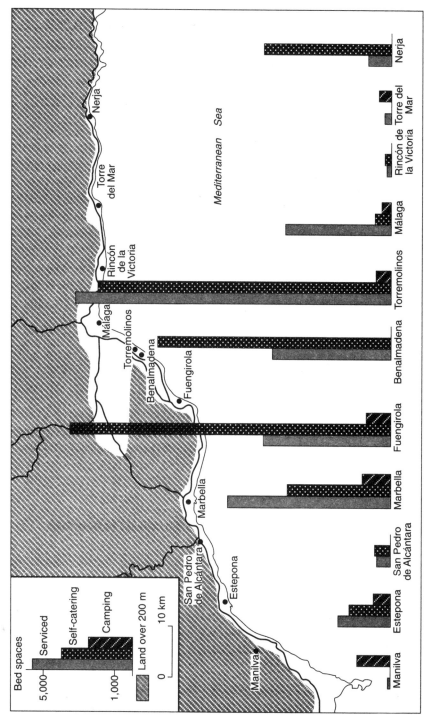

Source: Adapted from Patronato Provincial de Turismo de la Costa del Sol (1988).

Less positively, however, tourism's economic effect may also:

- increase dependence upon foreign investors and companies;
- introduce instabilities and weakness in labour markets;
- divert investment from other development areas.

Economic impacts are complex and notoriously difficult to isolate and measure, but we may be confident that there are significant spatial variations in effect according to:

- the geographic scale of development; i.e. international, national, regional or local;
- the initial volume of tourist expenditures, which will be primarily shaped by the number of visitors and their market segments. For example, effects differ between low-cost and luxury travel, or mass tourists and independent travellers.
- the size and maturity of the economy, which will affect particularly the ability to supply tourist requirements from within the economy rather than relying upon imports or foreign sources of investment;
- the levels of 'leakage' from the economy. Leakage represents the proportion of revenue which is lost through, for example, the need to import goods and services to sustain the tourism enterprise, or through the payment of profits and dividends to offshore owners or investors. In general, the larger and more developed an economy, the lower the levels of leakage and vice versa.

With these points in mind, let us now consider a cross-section of positive and negative impacts that have generally been associated with economic development based upon tourism.

## Positive impacts of tourism upon economic development

The first impact that we may note is the capacity for international forms of tourism to earn foreign currency and to influence, in a positive direction, a country's balance of payments account (this being the net difference between the value of exports and cost of imports). With a world tourism 'trade' currently valued at US$320 billion annually, the potential for tourism to influence the accumulation of wealth in particular regions is considerable.

Tourism is an example of what economists refer to as 'invisible' trade elements, meaning that such trade is not necessarily in tangible (and

**Table 4:1** *International balance of tourism trade: OECD members, 1994*

| Country | Receipts (US$ million) | Gross surplus (US$ million) | Gross deficit (US$ million) |
|---|---|---|---|
| USA | 60,406.0 | 16,844.0 | |
| France | 24,844.5 | 11,107.6 | |
| Italy | 23,754.3 | 11,669.5 | |
| Spain | 21,490.7 | 17,372.3 | |
| United Kingdom | 15,185.6 | | 7,011.1 |
| Austria | 13,151.6 | 3,752.2 | |
| Germany | 10,816.9 | | 31,531.3 |
| Switzerland | 7,629.5 | 1,254.5 | |
| Mexico | 6,318.0 | 957.0 | |
| Canada | 6,308.6 | | 3,125.1 |
| Australia | 5,902.6 | 1,933.4 | |
| Netherlands | 4,743.0 | | 4,496.4 |
| Belgium/Lux. | 4,666.5 | | 2,463.6 |
| Turkey | 4,359.0 | 3,493.0 | |
| Greece | 3,857.7 | 2,279.5 | |
| Portugal | 3,825.6 | 2,129.3 | |
| Japan | 3,464.4 | | 26,834.9 |
| Denmark | 3,175.0 | | 407.4 |
| Sweden | 2,838.0 | | 2,052.9 |
| Norway | 2,168.9 | | 1,773.7 |

Source: Organisation for Economic Co-operation and Development (1996).

hence easily measurable) flows of goods. However, we may gain an idea of the spatial patterning in national income and the gains and losses of foreign currency through tourism by comparing what a nation earns through foreign visitors' expenditure with what its own nationals expend when they themselves become tourists to another country. (This is sometimes referred to as the 'travel account'.) Table 4:1 sets out the balance of tourist trade in the top twenty OECD member countries, as measured by gross tourism receipts and according to whether countries are in surplus or deficit on their travel account.

Two points are worth noting. First, nations that are conspicuous generators of tourists, especially Germany, Japan and the United Kingdom, tend to be in 'deficit' on this particular form of measurement, Germany and Japan spectacularly so. However, Table 4:1 tells only a partial story since countries such as Germany and the United Kingdom are able to transform the contribution of tourism to their balance of payments – often from apparent deficit into surplus – through such pathways as ownership and control of tour companies, airlines, transport operators, international hotel chains and, less obviously, profits from insurance and banking services that support the international tourism industry. Second, there is a marked geographic pattern amongst the European nations, with a clear flow of foreign currency from the northern urban-industrial economies to the Alpine or Southern European economies, which, in the latter case, are generally more ruralised and less prosperous. (This illustrates a supplementary advantage that is often claimed to be associated with tourism development, namely that it can be a medium for redistribution of economic wealth from richer to poorer areas.)

A second advantage of tourism is its capacity to attract inward investment to finance capital projects. Although the industry is still dominated by small-scale local firms, the trend is towards greater levels of globalisation in the organisation of world tourism and the emergence of large-scale inter- and multinational operators, each capable of moving significant volumes of investment to new tourism destinations. These firms are distinctive not just because of the manner in which they have extended their horizontal linkages (where firms merge with, or take over, other firms operating in the same sector), but especially through the development of vertical linkages in which, for example, an airline purchases or develops its own travel company and takes on ownership of hotels. (The Grand Metropolitan group, for example, has interests in international hotels, holiday camps, travel agencies, package tours and restaurants.)

For developing nations, in particular, the role of foreign investment in initiating a tourism industry through, for example, hotel and resort construction can be an essential first step out of which an indigenous industry may eventually develop. Without foreign investment, start-up capital may not be found locally and although profits from foreign-owned firms will tend to leak out of the local economy, local taxation on visitors and their services may provide funding to assist in the formation of new indigenous firms and the development of key infrastructure (roads, water and power supplies) around which further expansion of tourism may then be based.

Third, tourism may play a role in processes of economic regeneration or in provision of new support for marginal economies through diversification. These powers have been evident for some time within rural economies. In Britain, in areas such as Wales or Devon and Cornwall, less profitable hill farm economies, or dairy farms where profits have been limited by EC production quotas, have been widely sustained by development of farm holidays and activities: fishing, riding, shooting, self-catering facilities, bed and breakfast businesses, caravanning and camping. More recently, however, tourism-based regeneration and diversification have been recognised in new forms of urban tourism. Active promotion of urban business tourism (conferences and conventions, etc.), sport and event-related tourism and development of new attractions centred around leisure shopping or industrial heritage has permitted places with no tradition of tourism to develop a new industry that has revitalised flagging local or regional economies. In the UK, for example, tourism and visiting to the port and industrial city of Liverpool generates an estimated £335 million annually and supports around 14,000 people in direct employment.

Fourth, tourism may promote development through the encouragement of new economic linkages and increase the gross domestic product (GDP) of an economy. Tourism's contribution to GDP will vary substantially according to the level of diversity and the extent of economic linkages within an economy. In a developed nation, tourism's contribution to GDP is usually quite small. For example, for the United Kingdom the share has typically been of the order of 1.0 to 2.0 per cent. In contrast, in emerging nations which lack economic diversity or which, perhaps through remoteness, have limited trading patterns, contribution of

**Table 4:2** *Tourism contributions to gross domestic product: selected countries, 1988*

| Country | Tourism as percentage of GDP | Country | Tourism as percentage of GDP |
| --- | --- | --- | --- |
| Bahamas | 53.0 | Austria | 7.9 |
| Barbados | 29.8 | Greece | 4.6 |
| Jamaica | 16.5 | Portugal | 5.8 |
| Kenya | 4.8 | Spain | 4.9 |
| Singapore | 9.7 | Sweden | 1.3 |
| Sri Lanka | 1.1 | Switzerland | 3.9 |
| Thailand | 5.4 | United Kingdom | 1.3 |

Source: Harrison (1994); Kaspar and Laesser (1993).

tourism to GDP can be substantial. Table 4:2 lists selected examples of estimated shares of GDP from tourism in a range of developed and developing nations.

The mechanisms by which tourism development may foster new-firm formation and development of new linkages are complex but in simplified form may be envisaged as shown in Figure 4:7. This simple model reflects a developing-world scenario in which at an initial stage, local provision is limited and the industry highly dependent upon overseas suppliers. After a time, numbers of tourism businesses increase and become more spatially spread, profits (or expectations of profit) filter more widely into the local economy and existing or newly formed local firms start to take up some of the supply market. Levels of foreign dependence therefore diminish as these local linkages emerge. Eventually, a mature stage is reached in which a broadly based local tourism economy has been formed with developed patterns of local supply and minimal dependence on foreign suppliers.

The potential contribution of tourism development to the wider formation of economic growth, inter-firm linkages and the generation of income is commonly assessed through what is termed the 'multiplier effect'. Multipliers attempt to measure the impact of tourist expenditure as it recirculates within a local economy. Tourist spending is initially introduced as direct payment for goods and services: accommodation, food, local transport, souvenir purchases, etc. In turn, the providers of these services respend a portion of their tourism receipts, for example in making their own purchases, in payment of wages to employees or in taxes to local government. These transactions form further flows of money and extend the indirect linkages of tourism well beyond the immediate core of tourism businesses. This cyclical process is reflected in the recognition of three levels of effect:

- a direct effect, which is the initial injection of revenue to the local economy by the tourist; for example, through payment of an hotel bill;
- an indirect effect, which is represented by a second round of spending by the recipients of initial expenditures in purchasing the goods and services demanded by the tourist; for example, purchase by the hotelier of local supplies for the hotel restaurant;
- an induced effect, which is further spending by the beneficiaries of the direct and indirect effects on goods and services for their own consumption; for example, the purchase of clothing by the hotel waiter.

**Figure 4:7** *Tourism development and the formation of economic linkages*

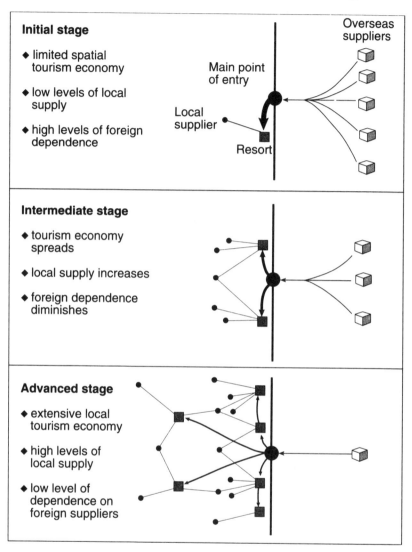

Source: Developed from Lungren, J.O.J. (1973) 'Tourist impact/island entrepreneurship in the Caribbean'. Paper presented to the Conference of Latin American Geographers, cited in Shaw and Williams (1994).

By convention, multipliers are expressed as a ratio in which the expected increase in income associated with a unit of currency is stated. Thus a multiplier of 1.35 would indicate that for every $1 spent, a further $0.35 dollars is generated by indirect and induced effects. However, the scale of the multiplier effect will vary – dependent upon the level of development within the economy, the type of tourism and the extent to which the local economy can supply the tourism industry from its own resources and, thereby, the extent to which the leakage effects may be minimised.

Fifth, tourism can be a significant source of employment, both direct employment within tourism businesses (e.g. hotels) and indirect employment in enterprises that benefit from tourism (e.g. general retailing). In comparison with many modern industries, tourism retains a relatively high demand for labour, based particularly around service work in hotels, restaurants, bars, retailing and local transportation. Smaller numbers work in the travel industry – within agencies, as couriers and guides, or in tourist information services, whilst smaller numbers still exercise managerial roles within the industry.

**Figure 4:8 Structure of the tourism labour market**

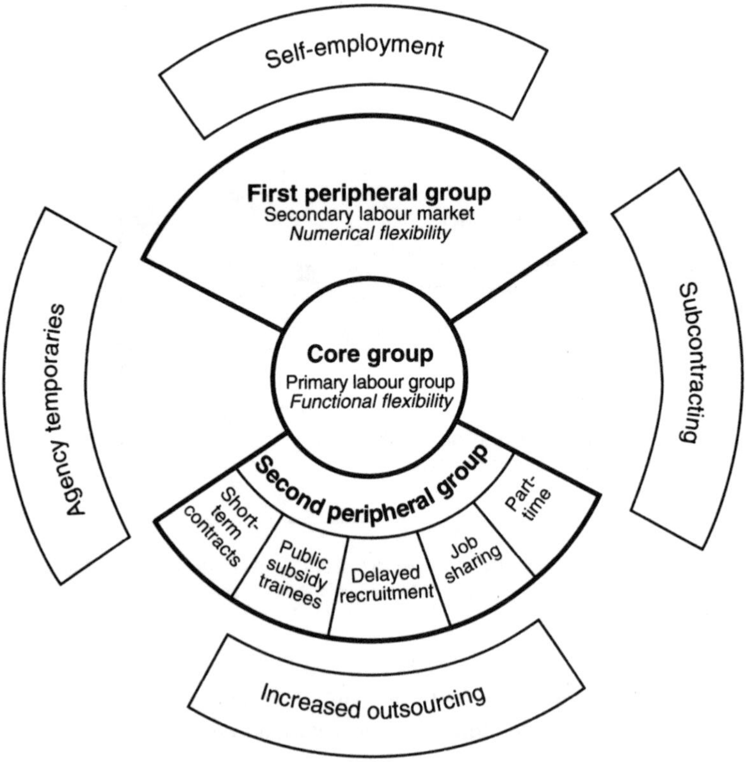

Source: Shaw and Williams (1994).

The structure of tourism employment has been usefully summarised in diagrammatic form (see Figure 4:8). Normally, tourism labour markets are centred around a relatively small core group of permanent, skilled managers and workers that form a primary labour source that is capable of a range of tasks (i.e. it is functionally flexible). Alongside the core are much larger secondary and tertiary groups that are more likely to be composed of relatively low-skilled personnel with more limited capabilities (i.e. *functionally* inflexible), but probably working part-time and therefore in sectors that are *numerically* flexible in their size and composition. Flexibility in the secondary labour market typically extends to importation of labour, and hence employment migration is often a distinctive geographic dimension in tourism economics. These structural characteristics are important since it means that major elements in tourism labour forces may be formed relatively quickly, with only modest levels of training and, equally, adjusted to reflect fluctuations in the market. From the perspective of developers and employers, at least, these represent considerable advantages.

## Negative impacts of tourism upon economic development

Although tourism development has a number of powerful attractions, there are also significant negative impacts. First, the industry is subject to instabilities, and in many tourism regions, climatic constraints produce a pronounced seasonal effect. Figure 4:9 illustrates the seasonality of tourism for a number of destinations, and from an economic perspective points to the problems of having facilities under-utilised or even closed (and therefore entirely unproductive) for parts of the season. Cutting across such seasonal patterning are more unpredictable fluctuations in demand within the industry. Tourism demand patterns are highly responsive to a number of potentially disruptive influences, including:

- economic recession in generating countries;
- changes in the price of holidays consequent upon fluctuations in international monetary exchange rates or price wars within the travel industry;
- changes in costs of transportation, reflecting particularly changes in oil prices and associated costs of aviation fuel;
- short- or medium-term economic and political instability in destination areas;
- warfare and civil unrest;

**Figure 4:9** *Seasonal patterns of international tourist arrivals in selected countries*

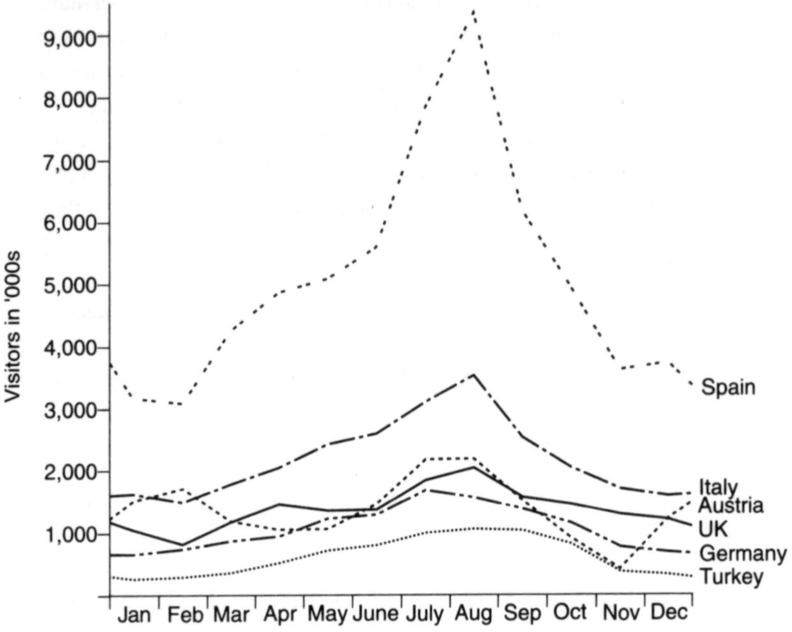

Source: Economics Intelligence Unit, *International Tourism Report* (various issues).

- negative images stemming from a range of potential problems at destinations, including levels of crime, incidence of illness and epidemics, or even simple decline in fashionability.

Thus, the oil crisis of 1974 or the recession of the early 1980s, civil unrest in Northern Ireland, attacks on tourists in places as different as Cairo or Miami, war in the Persian Gulf and in the former Yugoslavia, and the increasingly poor image of cheap Spanish resorts have all exerted effects upon tourism patterns at a range of geographic scales and led to the broad conclusion that tourism is often an unstable industry around which to base economic growth.

Second, in Third World nations in particular, tourism development may increase levels of economic dependence upon foreign companies and investors. Ideally, foreign investment provides a catalyst to growth that will foster the subsequent formation of local enterprise, yet in many emerging nations foreign ownership continues to dominate tourism industries, prompting extensive leakage of revenue and minimising the local economic gain. For example, amongst some of the Pacific microstates that are popular with Australian and New Zealand tourists, local ownership is minimal. On the Cook Islands, tourism businesses belonging to local people receive only 17 per cent of tourist expenditure, whilst in Vanuatu over 90 per cent of expenditure is gathered by foreign-owned companies.

Third, presumed capacities of tourism to generate regional development, redistribute wealth and benefit local economies have been questioned. Tourism has been widely associated with localised inflation of the price of land, labour costs and prices of goods in the shops, whilst studies of tourism destinations as diverse as the UK and Malaysia suggest that rather than diffusing economic gain into less wealthy, peripheral regions, development tends to refocus on areas of existing development. In Britain, traditional domestic tourism regions such as Devon and Cornwall have seen a loss of business with domestic holidaymakers flocking abroad. Ideally, such losses would be counterbalanced by new flows of foreign tourists into the same regions, yet the primary focus for foreign visitors to Britain is London, and only tiny numbers of overseas visitors venture as far as Cornwall. So at a regional level, the gains in foreign tourism in one locality are not compensating for losses in the domestic market, thereby sustaining rather than eroding regional disparities.

Fourth, questions have been raised over the role of tourism in generating local employment. Tourism work has been widely characterised as:

- low-paid;
- menial and unskilled;
- part-time and seasonal;
- over-dependent upon female labour.

Whilst such stereotyping over-simplifies a complex labour market and disregards the presence of a core of employees who fit none of these categories (see Figure 4:8), many tourism jobs do suffer from some (or all) of these characteristics. Such problems are particularly acute in cases of foreign development of tourism. Studies of tourism employment in the Caribbean and Africa, for example, show a recurring pattern with local labour placed into the low-pay, low-skill jobs, whilst positions with responsibility, higher earnings and prospects for advancement tend to go to foreign workers who possess appropriate skills and training. Even so, the allure of working in tourism and the enhanced working conditions that may prevail in comparison to other sectors (such as farming) can create imbalances in local workforces and create labour shortage in other sectors.

Lastly, any evaluation of the economic potential of tourism development needs to take some account of negative 'externalities'. These are the wider 'costs' that are attached to tourism development and which, whilst not always quantifiable in precise economic terms, nevertheless have an economic dimension. Traffic congestion in resort areas, over-use of water supply and sewerage systems and the pollution that may result, tourist-related crime – all have attached, though often unseen, costs that need to be recognised and entered into the overall economic balance.

To conclude this section, Box 4:4 offers an outline comparison of tourism and economic development in two developing nations which highlight several of the wider impacts discussed above.

---

## Box 4:4

### *Tourism and economic development in Tunisia and The Gambia*

Tunisia and The Gambia represent two examples of tourism development within the context of emerging nations and illustrate many of the advantages and disadvantages of tourism in an economy. Of the two, Tunisia is the more established destination, and benefits from its low costs, proximity to the European air charter market and the attraction of a Mediterranean coastline. Tourism in Tunisia has also been actively promoted and planned by successive governments, which have invested substantially in infrastructure, especially hotels. Between 1970 and 1992, the number of establishments rose from 212 to over 550, whilst available bed spaces increased even more rapidly from 34,000 to around 135,000 in 1992. Nearly 2 million European tourists entered the country in 1992, in addition to 1.5 million visitors from North Africa. However, numbers have fluctuated in response to political stability in the region, and the attraction of high-spending American tourists has proven particularly difficult in the face of increasing Muslim fundamentalism and anti-American attitudes in parts of North Africa and the Middle East.

In contrast, tourism to The Gambia (a much smaller West African state, set largely within the territory of a larger neighbour – Senegal) is modest in scale and has been less actively developed, with only minimal state investment; in 1991, foreign visitors to The Gambia numbered just over 100,000. Whereas Tunisia possesses several tourist zones, in The Gambia development is focused in the relatively small Bakau–Banjul development area, where the Gambia River enters the Atlantic Ocean.

Even so, the economic impact of tourism in The Gambia is not insignificant. Estimates suggest that tourism contributes a net income of US$ 25 million annually, lies second only to agriculture in its contribution to GDP and employs (directly and indirectly) over 7,000 people. In the larger Tunisian tourist industry, 20 per cent of foreign trade was accounted for by tourism and over US$ 900 million earned in foreign exchange, recovering nearly 45 per cent of the trade deficit on other sectors. Some 54,000 Tunisians are directly employed in tourism.

Because of heavy state investment in Tunisian tourism, a substantial share of tourism receipts accrue to the local economy, but in The Gambia, high levels of foreign ownership and the dominance of pre-paid package tours within the Gambian market erode apparent gains. The need for food imports to match European tastes is an additional requirement, and because of limited development of attractions outside resort areas, secondary expenditures in restaurants, shops and visitor attractions are minimal.

Employment patterns in both Tunisia and The Gambia reveal the common problems of a concentration of local labour into the low-skill, low-pay sectors – chambermaids, bar stewards, waiters, kitchen staff, etc. – whilst the middle and senior managerial levels are typically occupied by expatriates who bring the right skills and experience. Tunisia has established a number of training schools in an attempt to increase the proportion of local people who are qualified to work in the industry, and The Gambia clearly needs to do the same. Both countries also have to address problems of tourism attracting labour away from agriculture and other sectors, thereby creating imbalances in the labour supply.

Source: Dieke (1994); Poirier (1995).

# Summary

Processes of physical and economic development are the most visible ways in which tourism affects host areas. This chapter initially defines the primary factors that will shape patterns of tourism development and shows how they may combine to produce spatially contrasting forms. However, such developments not only alter the physical environments of destinations but also exert a range of economic effects too. These will vary from place to place, depending upon levels of local economic development, but could include a range of impacts upon balance of payments accounts, national and regional economic growth, and the creation of employment. Unfortunately, the instabilities of tourism that make it vulnerable to a range of influences (for example, exchange rate or oil price fluctuations; political crises; changes in fashion) mean the industry is not always able to provide a firm basis for economic development. For Third World countries, tourism may increase levels of foreign dependence, and in many contexts the quality of employment that the industry creates is low.

## Discussion questions

1 What are the principal elements that are needed to secure the physical development of a tourism destination?
2 How do variations in local conditions produce contrasting spatial patterns of tourism development?
3 Using examples of both established and emerging resorts, examine the validity of the models of resort structures provided in Figures 4:3 and 4:4.
4 What are the main strengths and weaknesses of tourism as a means of economic development?
5 Evaluate the potential of tourism as a source of local employment.

## Further reading

The physical development of resorts and tourism zones is given comprehensive coverage in:
Pearce, D.G. (1989) *Tourism Development*, Harlow: Longman.

For a broad discussion of the economic impacts of tourism see:
Mathieson, A. and Wall, G. (1982) *Tourism: Economic, Physical and Social Impacts*, Harlow: Longman.

and their recently revised edition of the same text:
—— (1997) *Tourism: Change, Impacts and Opportunities*, Harlow: Longman.

There is a convenient summary discussion in:
Ryan, C. (1991) *Recreational Tourism: A Social Science Perspective*, London: Routledge.

A useful examination of the economics of tourism in Europe which includes a great deal of case study material is provided by:
Williams, A.M. and Shaw, G. (eds) (1988) *Tourism and Economic Development: Western European Experiences*, London: Belhaven.

General discussions of tourism economics and development in a range of settings in the Third World are provided in:
Harrison, D. (ed.) (1994) *Tourism and the Less Developed Countries*, London: Belhaven/John Wiley.
Hitchcock, M., King, V.T. and Parnwell, M.J.G. (eds) (1993) *Tourism in South-East Asia*, London: Routledge.
Lea, J. (1988) *Tourism and Development in the Third World*, London: Routledge.

Good case studies of tourism and economic development are found in:

Dieke, P.U.C. (1994) 'The political economy of tourism in the Gambia', *Review of African Political Economy* No. 62: 611–627.

Lockhart, D., Drakakis-Smith, D. and Schembri, J. (eds) (1993) *The Development Process in Small Island States*, London: Routledge.

Milne, S. (1992) 'Tourism and development in South Pacific microstates', *Annals of Tourism Research*, Vol. 19: 191–212.

Oppermann, M. (1992) 'International tourism and regional development in Malaysia', *Tijdschrift voor Economische en Sociale Geografie*, Vol. 83 No. 3: 226–233.

Poirier, R.A. (1995) 'Tourism and development in Tunisia', *Annals of Tourism Research*, Vol. 22 No. 1: 157–171.

Tourism employment is examined in many sources, including:

Choy, D. (1995) 'The quality of tourism employment', *Tourism Management*, Vol. 16 No. 2: 129–137.

Shaw, G. and Williams, A.M. (1994) *Critical Issues in Tourism: A Geographical Perspective*, Oxford: Blackwell.

Urry, J. (1990) *The Tourist Gaze: Leisure and Travel in Contemporary Societies*, London: Sage.

# 5 ▸ Sustainable tourism? The environmental consequences of tourism development

The relationship between the environment and many forms of tourism is fundamental. From the earliest times, the enjoyment of 'environments' – whether defined in physical or in socio-cultural terms – has had a major impact in shaping a succession of tourism geographies. As public tastes for different kinds of leisure environment have developed through time – for example, through the formation of resorts or the changing preferences for scenic landscapes in the nineteenth century; or the quest for amenable climates or the attraction of historic heritage in the twentieth century – so new spatial patterns of interaction between people and environments have been formed.

However, tourism–environment relationships are not just fundamental, but also highly complex. There is a mutual dependence between the two that has often been described as 'symbiotic'. In simple terms this means that since tourism benefits from being located in good-quality environments, those same environments ought to benefit widely from measures of protection aimed at maintaining their value as tourist resources. In England and Wales, for example, the designation of national parks came about partly because these high-quality environments were seen as potentially valuable areas for tourism and the argument for their conservation was strengthened accordingly. Similarly, there is no doubt that the cause of wildlife preservation in East Africa has been assisted by the parallel increase in the appeal of safari holidays to the same region.

As tourism has expanded in the post-1945 period (both in scale and into new destinations) there have, however, emerged very real signs that the nature of that symbiosis has become unbalanced. Tourism, far from being a force for enhancement and protection of the environment, actually has shown itself to be a major generator of environmental problems with considerable capacity to destroy the resources upon which it depends. Consequently, more attention is now being focused upon understanding the environmental impacts of tourism and ways of producing more sustainable forms of tourism development that maintain, rather than degrade, key resources.

The complex character of tourism–environment relationships is deepened further by the diverse nature of those impacts and the inconsistencies through time and space in their causes and effects. It is also true that the effects of tourism upon the environment are partial, and one of the practical difficulties in studying those impacts is to disentangle tourist influences from other agencies of change that may be working on the same environment. So, for example, beach and inshore water pollution in the Bay of Naples will be partly attributable to the presence of tourists but will also be a product of the activities of local populations, of farming and of industries that discharge their waste into the Mediterranean.

The diversity of environmental impacts of tourism and the seriousness of the problem vary geographically for a number of reasons. First, we need to take account of the nature of tourism and its associated scales of effect. Impact studies often make the erroneous assumption that tourism is a homogeneous activity exerting consistent effects, but, as we have seen in Chapter 1, there are many different forms of tourism and types of tourist. The mass tourists who flock in their millions to the Spanish Mediterranean create a much broader and potentially more serious range of impacts than will small groups of explorers trekking in Nepal, although paradoxically, where mass forms of tourism are well planned and properly resourced, the environmental consequences may actually be less than those created by small numbers of people visiting locations that are quite unprepared for the tourist. For example, depletion of local supplies of fuel wood and major problems of littering have been widely reported along the main trails through the Himalayan zone in Nepal.

Second, it is important to take account of the temporal dimensions. In many parts of the world, tourism is a seasonal activity that exerts pressures on the environment for part of the year but allows fallow periods in which recovery is possible. So, there may be short-term/ temporary impacts upon the environment that may be largely coincident with the tourist season (such as air pollution from visitor traffic) or, more seriously, long-term/permanent effects where environmental capacities have been breached and irreversible changes set in motion (for example, reductions in the level of biodiversity through visitor trampling of vegetation).

Third, diversity of impacts stems from the nature of the destination. Some environments (for instance, urban resorts) can sustain very high levels of visiting because their built infrastructure makes them relatively resilient or because they possess organisational structures (such as planning frameworks) that allow for effective provision for visitors. In contrast, other places are much less robust, and it is perhaps unfortunate that a great deal of tourist activity is drawn (by tastes, preferences and habits) to far more fragile places. Coasts and mountain environments are popular tourist destinations that are often ecologically vulnerable, and even non-natural resources can suffer. Historic sites, in particular, may be adversely affected by tourist presence, and in recent years attractions such as Stonehenge, the Parthenon in Athens and the tomb of Tutankhamen in Egypt have all been subjected to partial or total closure to visitors because of negative environmental effects.

In exploring the environmental impacts of tourism, it is helpful to adopt a holistic approach to the subject. Environments, whether defined as physical, economic or social entities, are usually complex systems in which there are inter-relationships that extend the final effects of change well beyond the initial cause. Impact often has a cumulative dimension in which secondary processes reinforce and develop the consequences of change in unpredictable ways, so treating individual problems in isolation ignores the likelihood that there is a composite impact that may be greater than the sum of the individual parts. The effects of trampling of ground by tourists are a good example of this problem (see Figure 5:1).

A second advantage of a holistic approach is that it encourages us to work towards a balanced view of tourism–environment relationships. The temptation is to focus upon the many obvious examples of negative and detrimental impacts that tourism may exert, but, as the concept of a symbiotic relationship makes clear, there are positive effects too. These might be represented in the fostering of positive attitudes towards environmental protection/enhancement or might be reflected more

**Figure 5:1** *Effects of trampling at tourism sites*

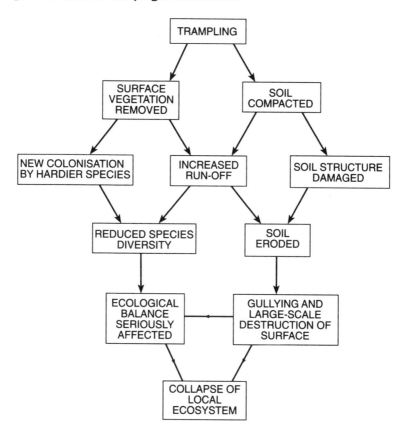

practically in actual investment in environmental improvement that restores localities for resident populations as well as providing support for tourism.

The third advantage of a holistic approach is that it recognises the breadth (some might say the imprecision) of the term 'environment' and the fact that different types of impact are likely to be present. As is perhaps implicit in the preceding discussion, the term can embrace a diversity of contexts – physical ecosystems; built environments; economic, social, cultural or political environments – and tourism has the potential to influence all of these, in varying degrees. For the purposes of this present discussion, the economic and socio-cultural impacts are discussed in other parts of the book (Chapters 4 and 7 respectively). So for the remainder of this chapter, the focus falls upon the influences that tourism may have upon physical environments, ecosystems and the built environment, together with a consideration of ways in which symbiotic relationships between tourism and the environment may be sustained.

## Environmental impacts of tourism: a review

Table 5:1 attempts to summarise a representative cross-section of positive and negative effects that tourism may have upon physical environments and proposes five key headings under which tourism effects may be grouped.

Under the first heading, *biodiversity*, are located a number of effects that broadly impact upon the flora and fauna of a host region. The 'balance' of influence here leans strongly towards the group of negative impacts, for whilst tourist demands have occasionally been partly or fully responsible for programmes aimed at establishing zones of conservation in which wildlife and their natural ecosystems are protected (for example, in the national parks in Kenya and Tanzania or on the Great Barrier Reef in Australia), the more commonplace patterns are associated with damage.

As Table 5:1 indicates, such damage may occur in varying forms. Most widely, processes of tourism development (construction of hotels and apartments, new roads, new attractions, etc.) can result in a direct loss of habitats. In the Alps, extensive clearance of forests to develop ski-fields and the loss of Alpine meadows with particularly rich stocks of wild flowers to new hotel and chalet construction has significantly altered ecological balances and, in the case of deforestation, greatly increased risks associated with landslides and snow avalanches.

At a more localised scale, other impacts become apparent. Destruction of vegetation at popular visitor locations through trampling or the passage of wheeled vehicles is a common problem. Typically, trampling causes more fragile species to disappear and to be replaced either by bare ground or, where regeneration of vegetation is possible, by more resilient species. The overall effect of such change is normally to reduce species diversity and the incidence of rare plants which, in turn, may impact upon the local composition of insect populations, insectivorous birds and possibly small mammals for which plant and insect populations are key elements in a food chain.

Larger animals may be affected in different ways by tourism, even within environments that are protected. The increasing popularity of safari

**Table 5:1 'Balance sheet' of environmental impacts of tourism**

| Area of effect | Negative impacts | Positive impacts |
|---|---|---|
| Biodiversity | Disruption of breeding/feeding patterns<br>Killing of animals for leisure (hunting) or to supply souvenir trade<br>Loss of habitats and change in species composition<br>Destruction of vegetation | Encouragement to conserve animals as attractions<br>Establishment of protected or conserved areas to meet tourist demands |
| Erosion and physical damage | Soil erosion<br>Damage to sites through trampling<br>Overloading of key infrastructure (e.g. water supply networks) | Tourism revenue to finance ground repair and site restoration<br>Improvement to infrastructure prompted by tourist demand |
| Pollution | Water pollution through sewage or fuel spillage and rubbish from pleasure boats<br>Air pollution (e.g. vehicle emissions)<br>Noise pollution (e.g. from vehicles or tourist attractions: bars, discos, etc.)<br>Littering | Cleaning programmes to protect the attractiveness of location to tourists |
| Resource base | Depletion of ground and surface water<br>Diversion of water supply to meet tourist needs (e.g. golf courses or pools)<br>Depletion of local fuel sources<br>Depletion of local building-material sources | Development of new/improved sources of supply |
| Visual/structural change | Land transfers to tourism (e.g. from farming)<br>Detrimental visual impact on natural and non-natural landscapes through tourism development<br>Introduction of new architectural styles<br>Changes in (urban) functions<br>Physical expansion of built-up areas | New uses for marginal or unproductive lands<br>Landscape improvement (e.g. to clear urban dereliction)<br>Regeneration and/or modernisation of built environment<br>Reuse of disused buildings |

Source: Adapted from Hunter and Green (1995).

holidays has become a problem in African national parks where the close attention of tourists in vehicles has been held to account for disruption to feeding and breeding patterns of animals and, in some cases, their eventual migration to remoter areas. (Box 5:1 provides an example of how quite innocent actions by tourists in one popular location impacts upon a particular species: the Mediterranean loggerhead turtle.) Nor has legal protection necessarily saved some animals and plants from decimation by collectors and tourists. Hunting of animals as a leisure pastime is still widely practised, and poaching to supply a black-market trade in animal souvenirs and trophies is commonplace in Africa, parts of the Mediterranean, the Caribbean and the South Pacific.

The impacts of tourism upon the diversity of flora and fauna link with the second area of concern, *erosion and physical damage*, and this illustrates how environmental problems tend to be interlinked. Erosion is typically the result of trampling by visitors' feet, and, whilst footpaths and natural locations are the most likely places for such problems to occur, extreme weight of numbers can lead to damage to the built environment. The Parthenon in Athens, for example, not only is under attack from airborne pollutants but also is being eroded by the shoes of millions of visitors. However, in such situations, tourism can have positive impacts, for although the activity may be a major cause of problems, revenue generated by visitors may also be a key source of funding for wider programmes of environmental restoration.

A more common problem is soil erosion, and Figure 5:1 attempts to show how the systematic manner in which the environment operates actually transmits the initial impact of trampling to produce a series of secondary effects which may eventually exert profound changes upon local ecosystems, leading to extensive damage. Localised examples of such damage can be spectacular. In north Wales, popular tourist trails to the summit of Snowdon now commonly reveal eroded ground that may extend to 9 m in width, whilst localised incidence of soil erosion and gullying has lowered path levels by nearly 2 m in a little over twenty years.

The environmental impacts of which the tourist is probably most aware are those associated with *pollution*, particularly the pollution of water. With so much tourism centred in or around water resources, pollution of water is a major concern. Poor-quality water may devalue the aesthetic appeal of a location and be a source of water-borne diseases such as gastro-enteritis, hepatitis, dysentery and typhoid. Visible water pollutants

# Box 5:1

## *Impact of tourism on wildlife: the example of the loggerhead turtle*

The Greek island of Zakynthos contains the most important nesting area in the Mediterranean for the loggerhead turtle, a species whose main habitat is the shallow inshore waters that also attract the tourist. Monitoring of turtle populations since 1979 (when the species was formally recognised as 'endangered') has shown a persistent decline in numbers, and whilst the turtles are vulnerable to several natural hazards that include climatic fluctuations and a rather wide range of natural predators, the development of tourism on Zakynthos has emerged as one of the greatest risks to the long-term survival of the species.

The turtles nest during the height of the summer tourist season, laying eggs in buried chambers in the beach, some 10 to 15 m from the water's edge. This unhappy set of coincidences, although quite inadvertent, directly disrupts breeding in several ways:

- some nesting sites have been lost to beach developments and improvements (such as tree planting to shade tourists);
- nesting females and young hatchlings (which are positively phototactic – i.e. attracted by light), rather than heading instinctively to the sea, may be disoriented by lights from beach-front bars and cafes, becoming stranded far from the sea;
- noise is also a source of disorientation and confusion;
- vehicular traffic on the beaches compacts the sand, reduces essential oxygen levels within nest chambers and may lead to collapse of nests;
- pollution of the water leads to fatalities as turtles consume plastic bags and food packaging, mistaking these items for natural foods such as jellyfish.

Initiatives by the Greek government to limit the impacts of tourism upon the turtles have so far proven only partially successful. Attempts to limit developments and constrain activities have drawn opposition from local residents who are dependent upon tourism, whilst visitors – although expressing broadly based concerns for the welfare of the animals – also show varying levels of disregard for restrictions aimed at protecting nest sites from intrusion. The authors of the study conclude that a primary goal of policies aimed at protecting the animals must therefore be educative in nature, seeking to alter the attitudes, values and behaviour of both the providers and the consumers of tourism on Zakynthos.

The study shows clearly the incidental manner in which tourism can disrupt wildlife. The animals are neither actively hunted nor pursued by tourists with cameras and most of the movement of turtles takes place at night and is largely unseen. Yet routine behaviour by tourists – pursued without any disruptive intention whatsoever – is nevertheless having seriously deleterious effects upon the species.

Source: Prunier *et al.* (1993).

(sewage, organic and inorganic rubbish, fuel oil from boats, etc.) will also be routinely deposited by wave action onto beaches and shorelines, leading to direct contamination, noxious smells and visually unpleasant scenes.

Pollution of water also has a number of direct effects upon plant and animal communities. Reduced levels of dissolved oxygen and increased sedimentation of polluted water diminish species diversity, encouraging rampant growth of some plants (e.g. various forms of seaweed) whilst discouraging less robust species. In some cases, such changes have eventually impacted upon tourists. In parts of the Mediterranean, and particularly the Adriatic Sea, the disposal of poorly treated sewage (supplemented by seepage of agricultural fertilisers into watercourses that feed into the sea) has created localised eutrophication of the water. (Eutrophication is a process of nutrient enrichment.) This has led directly to formation of unsightly and malodorous algal blooms that coat inshore waters during the summer months, reducing the attractiveness of the environment and depressing demand for holidays in the vicinity (see Box 5:2).

Water pollution is especially commonplace in areas of mass tourism where the industry has developed at a pace that is faster than local infrastructures have been able to match (for example, the Spanish Mediterranean coast), but even in long-established tourism locations, where local water treatment and cleansing services ought to be adjusted to local needs, water pollution is still commonplace. In 1996 the UK Environment Agency reported that 11 per cent of beaches in England and Wales failed to comply with EU minimum standards governing faecal contamination of bathing waters. In some regions (such as the North West of England – which covers some of the most popular holiday beaches at resorts such as Blackpool and Southport) as many as 40 per cent of the bathing waters were below EU targets and not a single beach merited the coveted EU 'Blue Flag' for beach cleanliness.

Alongside water pollution, tourism is also associated with air pollution and, less obviously, noise pollution. Pollution due to noise is usually highly localised, centring upon entertainment districts in popular resorts, airports and routeways that carry heavy volumes of tourist traffic. However, the dependence of tourism upon travel means that chemical pollution of the atmosphere by vehicle exhaust fumes is more widespread and, given the natural workings of the atmosphere, more likely to travel beyond the region in which the problem is generated. Nitrogen oxides,

## Box 5:2

### *Water quality and tourism: the case of Rimini*

One of the most recent examples of the mutual dependence between tourism and environmental quality has been the impact of deteriorating water conditions on tourism to the Italian resort of Rimini on the Adriatic. The River Po and its tributaries transport considerable volumes of urban, agricultural and industrial wastes into the Adriatic, to which is added waste from the coastal resorts themselves. The limited tidal range in the Adriatic has meant that pollutants have gradually accumulated, leading to localised eutrophication of the water and the formation, during the summer months, of algal blooms and floating rafts of mucilage. The first manifestation, which was patchy in form, occurred during August 1988, but the algae bloom was far stronger and more extensive in 1989 and attracted widespread media coverage.

The attention of the media and the negative images of polluted, algae-strewn waters had an immediate impact upon tourism, with an estimated reduction by 1990 of between 50 and 60 per cent in organised and intermediate forms of international tourism, although the loss of tourists in traditional, local sectors was much less marked.

Faced with economic catastrophe, the initial responses – from the tourism industry at least – were to treat the symptoms rather than the causes. Some use was made of floating barriers to limit the incursion of the rafts of algae inshore, and mobile bathing pools were erected at some locations. Discounted prices were widely used to try to maintain a market share.

However, the long-term viability of resorts with eutrophication problems rests on a more fundamental understanding of water pollution and local environmental systems, and to that end the Emilia-Romagna regional authority has set up programmes aimed at:

- monitoring water conditions;
- undertaking research into coastal currents and sedimentation characteristics in relation to certain pollutants;
- obtaining better understanding of algal development processes.

The European Parliament, noting that eutrophication problems are not confined to the Adriatic but are also manifest in parts of the Baltic and North Seas, has focused attention on the need for more rigorous management and control of agricultural, domestic and industrial pollution as a basic set of causal factors linked to eutrophication.

The recent experience in the resort of Rimini illustrates well the fragile interdependence between tourism and the environment. The resort was already showing signs of deterioration through physical overdevelopment and associated reductions in quality, so the adverse publicity associated with poor water conditions simply acted as a catalyst for extensive relocation of tourists to other areas where environmental problems were not perceived to be present. The extent to which Rimini will be able to recover its position will depend very much upon its ability to resolve the problem of algal blooms and the associated pollution, together with positive marketing to counteract the negative image that the resort has now acquired.

Source: Becheri (1991).

lead and hydrocarbons in vehicle emissions not only threaten human health but also attack local vegetation and have been held to account for increased incidence of acid rain in popular localities. The St Gotthard Pass, which lies on one of the main routeways between Switzerland and Italy, is one location where atmospheric pollution from tourist traffic has been responsible for extensive damage to vegetation, including rare Alpine plants.

A fourth area of concern centres on tourism impacts upon the *resource base*. Whilst tourism may be an agency for the promotion of resource conservation measures, it will exert negative effects associated with depletion or diversion of key resources. The attraction of hot, dry climates for many forms of tourism creates particular demands for local water supplies, which may become depleted through excessive tourist consumption or be diverted to meet tourist needs for swimming pools or well-watered golf courses. In parts of the Mediterranean, tourist consumption of water is as much as six times the levels demanded by local people. Tourism may also be responsible for depletion of local supplies of fuel or perhaps building materials. Paradoxically, the removal of sand (for concrete) from beaches is not uncommon.

The final area of environmental impact concerns *visual and structural changes*, and it is here that there is perhaps the clearest balance between negative and positive impacts of tourism. The physical development of tourism will inevitably produce a series of environmental impacts. The natural and non-natural environment may be exposed to forms of 'visual' pollution prompted by new forms of architecture or styles of development. Land may be transferred from one sector (for example farming) to meet demands for hotel construction, new transport facilities, car parks or other elements of infrastructure. The built environment of tourism will also expand physically, whether in the form of accretions of growth on existing urban resorts, new centres of attraction or second homes in the countryside.

However, set against such potentially adverse changes, there are significant areas of benefit. First, tourist-sponsored improvements to infrastructure, whether in the form of enhanced communications, public utilities or private services, will have some beneficial effects for local residents too. Second, tourism may provide a new use for formerly unproductive and marginal land. The rural environments in central Wales, north-west Scotland and the west of Ireland, for example, have all been partly sustained by the development of rural tourism. Third, tourism to cities has helped to promote urban improvement strategies aimed at clearing dereliction. Examples include the programme of national and international garden festivals held in several British cities during the 1980s which took derelict industrial sites and created new tourist attractions out of the wasteland. In Britain, continental Europe, the USA and Canada, the regeneration through reuse of redundant areas – dockland and water frontages being favoured targets – has been a recurring theme in contemporary urban development.

## Towards a sustainable relationship between tourism and environment

The evident problems that surround tourism and the environment have led to the formulation of a range of management responses to the perceived difficulties. This has been mirrored both in the development of site-specific management techniques and also, more fundamentally, in strategies and approaches aimed at developing sustainable forms of tourism.

There are a number of tourism management techniques that have been widely applied in areas where protection of environments is a key consideration – for example, within designated national parks. These techniques normally focus upon:

● spatial zoning;
● spatial concentration or dispersal of tourists;
● restrictive entry or pricing.

Spatial zoning is an established land management strategy that aims to integrate tourism into environments by defining areas of land that have differing suitabilities or capacities for tourism. Hence zoning of land may be used to exclude tourists from primary conservation areas; to focus environmentally abrasive activities into locations that have been

**Figure 5:2   Spatial zoning strategies in the Peak District National Park, England**

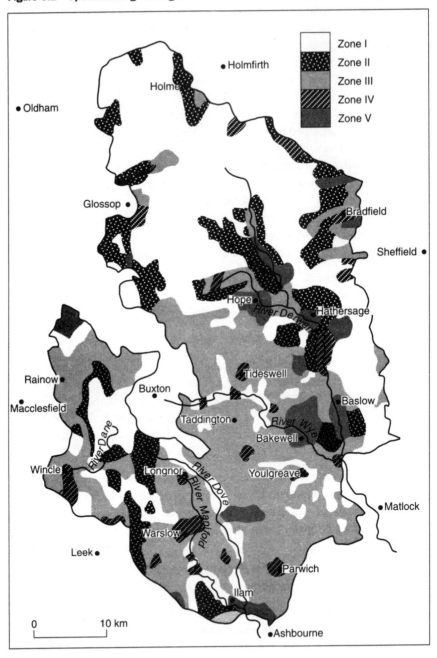

specially prepared for such events; or to focus general visitors into a limited number of locations where their needs may be met and their impacts contained and managed. Figure 5:2 shows a zoning policy developed within the Peak District National Park in northern England.

Zoning policies are often complemented by strategies for concentrating tourists into preferred sites (sometimes referred to by recreational planners as 'honeypots') or, where sites are under pressure, deflecting visitors to alternative destinations. Honeypots are commonly provided as interceptors – planned locations that attract the tourist by virtue of their promotion and on-site provision (e.g. information, refreshment, car parking, etc.) and which then effectively prevent the further penetration of tourists into more fragile environments that may lie beyond. (Commercial tourist attractions, tourist information and visitor centres, country parks and heritage sites are all examples of locations that can act as honeypots and assist in the wider environmental management of tourism.) In contrast, where conditions require a redistribution of tourist activity, devices such as planned scenic drives or tourist routes may have the desired effect of taking people away from environmental pressure points.

In some locations, regulation of environmental impacts of tourism is now being achieved via pricing policies and/or exclusions and controls. The nature and scope of such practice varies considerably from place to place. In the USA, for example, entry to many of the national parks is subject to payment of an entry toll, whereas in England and Wales, entry is free. However, policies of exclusion and control are commonplace and becoming more so through time as the pressures of tourism grow. The stated policy of the Dartmoor National Park authority, for example, is to deflect as much tourism development as is possible to the periphery of the park, in order to protect the open moorland environment that lies at its core. Within the park, visitors are encouraged (through patterns of access and planned provision) towards a relatively small number of higher-capacity sites, whilst movement of vehicles is subject to a park-wide traffic policy which both restricts and segregates vehicles to prescribed routes according to size and weight (Figure 5:3).

In some senses, however, these practical techniques for harmonising tourism and the environment are simply building-blocks that lead towards the much broader goal of securing sustainable forms of tourism development. Sustainable development is a concept that has entered the language in a diversity of contexts – population growth, natural resource

**Figure 5:3** *Traffic management strategies in Dartmoor National Park, England*

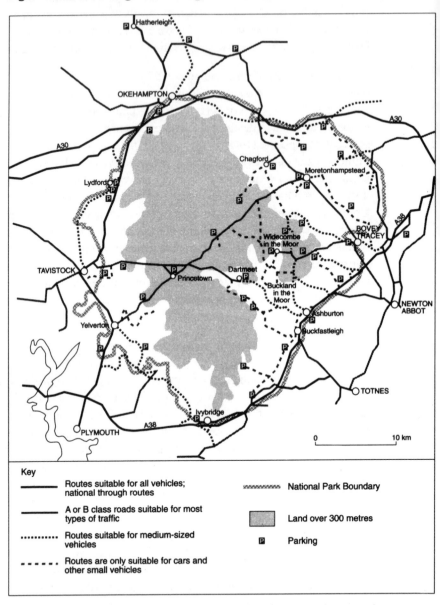

Key

| | |
|---|---|
| ———— | Routes suitable for all vehicles; national through routes |
| ———— | A or B class roads suitable for most types of traffic |
| ·········· | Routes suitable for medium-sized vehicles |
| - - - - - | Routes are only suitable for cars and other small vehicles |
| ∿∿∿∿∿ | National Park Boundary |
| ▨ | Land over 300 metres |
| P | Parking |

development, energy consumption and, not least, tourism – and advocates of sustainability argue its merits as the most effective, long-term resolution of a range of environmental and resource-related problems. But what does the term 'sustainable development' actually mean, both as a general principle and in the context of tourism?

## Sustainable tourism

The concept of sustainability has been defined by the World Commission on Environment and Development as 'development that meets the needs of the present without compromising the ability of future generations to meet their own needs'. When viewed in these terms, the relevance of sustainable forms of development to tourism is obvious, given that it is an industry with a high level of dependence upon 'environments' as a basic source of attraction but also one, as we have seen, with a considerable capacity to erode the long-term viability of those self-same environments. Tourism therefore needs to be involved in sustainable development.

However, the outwardly simple definition of sustainability cited above conceals much controversy and debate over who defines what is, or is not, sustainable and what sustainable development might therefore mean in practice. The concept implicitly recognises that there are basic human needs (e.g. food, clothing, shelter) that processes of development must match and that these needs are to be set alongside aspirations (e.g. to higher living standards, security and access to discretionary elements such as tourism) that it would be desirable to match. At the same time, however, there are environmental limitations that will ultimately regulate the levels to which development can actually proceed, and if principles of sustainability are also to embrace implicit notions of equity in access to resources and the benefits that they bring, then sustainability is likely to prove an elusive target in the absence of some radical shifts in attitudes and beliefs. For this reason, the concept of sustainability has acquired a diversity of interpretations ranging from, at one extreme, a 'zero-growth' view that argues that all forms of development are inherently unsustainable and should therefore be resisted, to very different perspectives that argue for growth-oriented resource management based around the presumed capacities of technology to solve environmental problems and secure a sustainable future.

For the further development of tourism (as for most areas where sustainability is an issue), a middle path between these extremes – one which manages growth within acknowledged resource conservation limits – is generally held to offer the best prospects. Sustainable tourism needs therefore to:

● ensure that renewable resources are not consumed at a rate that is faster than rates of natural replacement;

- maintain biological diversity;
- recognise and value the aesthetic appeal of environments;
- follow ethical principles that respect local cultures, livelihoods and customs;
- involve and consult local people in development processes;
- promote equity in the distribution of both the economic costs and the benefits of the activity amongst tourism developers and hosts.

## Approaches to the evaluation of environmental impacts and sustainable tourism

The central attributes of sustainable tourism may be mapped relatively easily, but the more practical difficulties of how to measure sustainable forms of development are less easily resolved. This particular challenge has focused attention onto alternative approaches to evaluating the environmental impacts of tourism; in particular, the concept of carrying capacity; the limits of acceptable change; and the use of environmental impact assessments.

The concept of carrying capacity is a well-established approach to attempting to understand the ability of tourist places to withstand use and is inherent in the notion of sustainability. It recognises that for any environment, whether natural or non-natural, there is a capacity (or level of use) which when exceeded is likely to promote varying levels of damage and/or be associated with reduced levels of visitor satisfaction. Carrying capacity has been visualised in several distinct ways; for example:

- as *physical* carrying capacity – which is normally viewed as a measure of absolute space, such as the number of spaces within a car park;
- as *ecological* capacity – which is the level of use that an environment can sustain before damage to the environment is experienced;
- as *perceptual* capacity – which is the level of crowding that a tourist will tolerate before he or she decides a location is too full and relocates elsewhere.

Although carrying capacity is easy to conceptualise, the value of the idea as a tool for measurement of impact is more limited. Ecological capacities are difficult to anticipate whilst perceptual carrying capacities, as personalised responses, are prone to variation both between and within individuals and tourist groups, depending very largely upon circumstance and motives. Capacity will also vary according to prevailing management practices whereby a site that is actively planned for tourism is likely to have a higher set of capacities than one that is not.

As a result of these limitations, alternative approaches to impact assessment have become more popular. The limits of acceptable change (LAC) technique was developed in the USA as a means of resolving development-related conflicts in conservation areas. The central features of the method (which is summarised as a set of key stages in Table 5:2) are:

- the establishment of an agreed set of criteria surrounding a proposed development;
- the representation of all interested parties within decision-making;
- the prescription of desired conditions and levels of change after development;
- the establishment of ongoing monitoring of change and implementation of agreed strategies to keep impacts of change within the established limits.

The LAC approach therefore embodies several key aspects of sustainable forms of tourism development. It recognises that change is an inevitable consequence of development but asserts that by the application of rational planning, overt recognition of environmental quality considerations and broad public consultation, sustainable forms of development may be realised. However, the approach does suffer practical weaknesses too. There are technical difficulties in agreeing and assessing qualitative aspects of tourism development and the process is dependent upon the existence of a structured planning system and sufficient resources in expertise and capital to operationalise the

**Table 5:2  *Key stages in the limits of acceptable change (LAC) process***

- Background review and evaluation of conditions and issues in development area
- Identification of likely changes and suitable indicators of change
- Survey of indicators of change to establish base conditions
- Specification of quality standards to be associated with development
- Prescription of desired conditions within zone(s) of development
- Agreement of management action to maintain quality
- Implementation, monitoring and review

Source: Adapted from Sidaway (1995).

monitoring and review stages. Hence the contexts in which LAC would be most beneficial – for example, in shaping tourism development in Third World nations – often prove the least suited to the technique in practical terms.

The same constraint is also true of the third possible approach to realising sustainable development, the use of environmental impact assessment (EIA). EIA is becoming a widely used method for evaluating possible environmental consequences of all forms of development and is potentially a valuable tool for translating sustainable principles into working practice. In particular, EIA provides a framework for informing decision-making processes that surround development, and a widening number of industries are routinely required (or advised) to undertake EIAs and produce written environmental impact statements (EISs). Table 5:3 summarises the four key principles to which most EIAs will conform.

**Table 5:3  *Key principles of environmental impact assessment***

- Assessments should identify the nature of the proposed and induced activities that are likely to be generated by the project
- Assessments should identify the elements of the environment that will be significantly affected
- Assessments will evaluate the nature and extent of initial impacts and those that are likely to be generated via secondary effects
- Assessments will propose management strategies to control impacts and ensure maximum benefits from the project

Source: Adapted from Hunter and Green (1995).

The methodologies of EIA are diverse and may embrace the use of key impact checklists, cartographic analysis of spatial impacts, simulation models or predictive techniques. Their strengths are that when properly integrated into the planning phases of a project, they should help developers anticipate environmental effects, enable more effective compliance with environmental standards and reduce need for subsequent (and expensive) revision to projects. The overall goal of a sustainable form of development is also a more achievable object when environmental impacts have been evaluated in advance. However, EIA has also attracted criticisms, which have included tendencies:

- to focus on physical and biological impacts rather than the wider range of environmental changes;

- to application on a project-specific basis and/or at the local geographic level, thereby overlooking wider linkages and effects;
- to require developed legislative and institutional frameworks in which to operate;
- to require a range of scientific and other data as a means of assessing likely impacts;
- to advocate technocratic solutions to environmental problems, which some advocates of sustainable development view as inappropriate.

Thus, as with LAC, EIA has practical limitations that will inhibit its application in many tourism development contexts, especially those that might benefit most from the technique.

The relative recency with which sustainable tourism has become an active issue means that practical cases that exemplify the application of sustainable principles and the methodologies outlined above are still not widespread, but Box 5:3 presents an outline summary of two contrasting cases that do show how sustainable forms of environmental management are being applied in tourism areas. In the first example, attempts to protect a rare ecosystem on the Great Barrier Reef in Australia are outlined, whilst in the second case, a successful programme of sustainable water management at the desert resort of Palm Springs, California, is described.

---

## Box 5:3

### Sustainable tourism in practice: Australia's Great Barrier Reef and Palm Springs, California

## The Great Barrier Reef, Australia

The Great Barrier Reef off the northern coast of Queensland, Australia, provides one of the largest systems of coral reefs in the world. A maze of some 600 islands, 300 cay (reef islands) and nearly 3,000 submerged reefs, the Great Barrier Reef region is home to 1,500 species of fish, about 350 types of coral, over 400 sponges and more than 4,000 molluscs. Apart from some small areas of development, the reef has been largely unaffected by human activity and remains in excellent condition.

However, in recent years there have been significant increases in the presence of tourists. Over a fifteen-year period up to around 1990, tourist bed spaces increased from 785 to over 2,000; the charter vessel fleet grew from 135 to 300 boats and numbers of registered speedboats from 15,000 to 24,000. In 1988, over 900,000 tourists visited the reef and the growing popularity of tourism to the area has begun to exert a range of environmental impacts. These have included physical destruction of reefs by trampling effects of divers standing on the coral and damage by boat anchors; localised water pollution from sewage and boat fuel; and removal of corals and specimen fish as souvenirs.

As a direct response to these problems, the Great Barrier Reef Marine Park (which at 350,000 km$^2$ is the largest protected marine area in the world) was set up by the Queensland Federal Government in 1975 to manage the conservation and use of the Reef. The Marine Park Authority has a broad remit that includes research, preparation and implementation of management plans, educational programmes and the regulation of commercial fishing.

The main strategy within the park management is a zoning system based around three primary categories:

- 'General use' zones (which cover about 80 per cent of the park) permit most activities, provided they are ecologically sustainable.
- 'National park' zones allow only activities that do not remove living resources.
- 'Preservation' zones permit only scientific research.

Within zones, specific local variations may also be enforced, particularly limitations on building.

Tourism (and its associated developments) may occur in all zones except the preservation zones, subject to the issue of permits. The factors considered in issuing permits will include the objectives of management within the zones in question; the size, extent and location of the use; access conditions; likely effects upon the environment in general and the ecosystem in particular; and likely effects upon resources and their conservation. Proponents of large-scale developments are also encouraged to conduct an EIA and to produce an EIS as a routine part of development applications. The tourists themselves are targeted via educational strategies, reinforced by local controls and prohibitions, aimed at encouraging responsible behaviours that help to conserve the marine environment.

## Palm Springs, California

The development of the desert resort of Palm Springs is a remarkable example of sustainable tourism in a difficult environment. Located in the Coachella Valley some 160 km south-east of Los Angeles, Palm Springs is a true desert region with mean July temperatures of 42°C and less than 75 mm of rain per year. However, the presence of springs fed from a substantial aquifer initially allowed the development of irrigated farming and, more recently, the growth of a fashionable tourist resort. More than 2 million visitors annually visit Palm Springs and, with over 200 hotels, 7,500 swimming pools and more than 80 golf courses, the Coachella Valley has become a major recreational environment within southern California.

The success of the resort has, however, depended entirely upon the sustainable management of its water supplies. The key to the development was the construction in 1948 of the Coachella branch of the All-American Canal, which transfers water from the Colorado River near Yuma to the valley. This new supply not only helped to recharge the main groundwater sources, which farming had already begun to diminish, but also created a surplus of water that permitted the expansion of the resort. Diverted water is trapped in intake basins from where it percolates into underground storage, and despite the increasing demand for water in Palm Springs, groundwater reserves have actually increased in recent years, rather than diminished.

To reinforce the effectiveness of the scheme for recharging groundwater, a wide-ranging programme for managing water supply and demand has been implemented. This has included:

- improved extraction techniques to maximise the potential of groundwater sources;
- improved application systems, including computer-controlled drip irrigation systems and metered supplies;
- increased charges to moderate demand;
- increased reuse of waste water.

The latter policy, in particular, has benefited the tourism industry as some 3 million gallons a day of cheap reclaimed water is distributed to parks, urban amenity spaces and, particularly, the resort's golf courses.

Sustainability has also been increased by complementary moves towards more water-efficient urban design and landscaping in Palm Springs. New golf courses are now encouraged to limit watered turf to only the essential parts of the course – mainly the greens, tees and key sections of fairway – whilst urban parks, hotels, civic buildings and even private residences have been persuaded to adopt the practice of low-water-use landscaping using desert plants and natural surfacing. Although extensive areas of lush, green ornamental space still adorn the resort, the wider use of desert-style landscaping has permitted a 10 per cent reduction in outdoor use of water in five years, even though parkland and amenity acreages have actually risen.

Source: Craik (1994); Pigram (1994).

## Sustainability and alternative forms of tourism

The examples of the Great Barrier Reef and the desert resort of Palm Springs demonstrate sustainability in the context of conventional forms of tourism, but, in concluding this chapter, one further question merits brief attention. How far do the so-called 'alternative' forms of tourism provide templates for sustainable tourism in general? There is perhaps a natural temptation to view the mass forms of packaged tourism as the least sustainable and the style of tourism that is most likely to bring widespread environmental change. In contrast, alternative forms of tourism (which are often characterised by their smaller scale, the involvement of local people, a preference for remoter areas and a predilection to place enjoyment of nature, landscape and cultures at the centre of the tourism experience) outwardly appear more in tune with principles of sustainability. Further, the alluring names that are commonly given to alternative forms of tourism – 'green tourism', 'eco-tourism', 'soft tourism', 'responsible tourism', even 'sustainable tourism' – tend to reinforce a popular belief that sustainability can only be equated with alternative tourism.

Such views do, however, need to be accorded considerable caution, for although the underlying philosophies of alternative tourism may strongly reflect the concept of sustainability, the experience of alternative tourism in a growing number of places suggests that such forms may be highly potent as agents of change and generators of impact. In fact, alternative tourism can be just as problematic, in development terms, as mass forms of tourism.

Several potential problems have been noted. First, alternative tourism usually penetrates far deeper into the personal lives of residents than the more aloof forms of mass tourism, with similarly enhanced capacities to generate a range of environmental, economic, social and cultural impacts. Second, lack of local expertise in catering for alternative tourists can mean that inappropriate practices are implemented and local resources over-exploited for short-term gain. Third, there is an evident risk that the alternative forms of tourism simply represent the pioneering stages in new practices of mass recreation. In this way, alternative tourism simply becomes a mechanism for constructing new geographies of travel and its associated impacts, centred on the exotic and the distant. There is ample evidence in high-street travel agencies that destinations that until quite recently were the domain of the alternative tourist – for example, the Himalayas, China, sub-Saharan Africa – are now being opened up to the package tourist, albeit at the luxury end of the market for the present.

Perhaps most fundamentally, alternative tourism – whilst perhaps embracing many principles of sustainability – does not in itself provide a model for sustainable forms of mass tourism. As several writers have noted, alternative tourism is not a replacement for mass tourism. It lacks the physical capacity, logistics and organisation to meet the growing levels of demand, it lacks the economic scale that has become so important to many national, regional and local economies, and the style of alternative tourism fails to match the tastes and preferences of many millions of holidaymakers and travellers world-wide.

So, whilst there are aspects of alternative tourism that certainly provide lessons in how to forge sustainable relationships between tourism and the environment, alternative tourism is not a natural (sustainable) replacement for the supposedly problematic mass forms of travel. Solutions to the problem of sustainability therefore need to be forged within the context of mass tourism, and that suggests that if the symbiotic relationship between tourism and the environment is to be maintained, careful management and planning of tourism development – whether guided by sustainable principles or not – must be a central component in the future growth of tourism. The role of planning in tourism forms the focus for the next chapter.

## Summary

Many forms of tourism are dependent upon the environment to provide both a context and a focus for tourist activity, yet those same activities have a marked capacity to devalue and, occasionally, destroy the environmental resources upon which tourism is based. Environmental effects of tourism are broadly experienced in impacts upon ecosystems, landscapes and the built environment, although specific impacts vary spatially – reflecting differences in the nature of the places that tourists visit, the levels and intensity of development, and the skills and expertise of resource managers. As the environmental problems associated with tourism have become more apparent, greater attention has been focused upon ways of producing sustainable patterns of development and alternative forms of tourism that produce fewer detrimental effects upon the tourist environment. However, truly sustainable tourism has often proven to be elusive, whilst there are evident risks that alternative tourism, in time, develops into mass forms of travel, with all the attendant problems that such practices tend to produce.

## Discussion questions

1  What are the main factors that will lead to spatial variation in the environmental impacts of tourism?
2  What do you understand by the concept of a 'symbiotic relationship' between tourism and the environment?

3  How do ecological systems transmit some environmental impacts of tourism well beyond their initial sources?
4  How far can conceptual tools such as carrying capacity, limits of acceptable change and environmental impact assessment actually help us to create sustainable forms of tourism?
5  Why should we treat with caution the popular assumption that alternative forms of tourism are intrinsically sustainable?

## Further reading

A good understanding of the environmental impacts of tourism is provided by:
Mathieson, A. and Wall, G. (1982) *Tourism: Economic, Physical and Social Impacts*, Harlow: Longman.
——— (1997) *Tourism: Change, Impacts and Opportunities*, Harlow: Longman.

whilst the following provides a convenient critique of links between tourism and sustainability:
Hunter, C. and Green, H. (1995) *Tourism and the Environment: A Sustainable Relationship?*, London: Routledge.

Other studies of sustainable development are to be found in:
Briguglio, L., Butler, R.W., Harrison, D. and Filho, W.L. (1996) *Sustainable Tourism in Islands and Small States: Case Studies*, London: Pinter.

Several essays on the themes of alternative forms of sustainable tourism (including ecotourism) are to be found in:
Smith, V.L. and Eadington, W.R. (eds) (1994) *Tourism Alternatives: Potentials and Problems in the Development of Tourism*, London: John Wiley.
Theobald, W. (ed.) (1994) *Global Tourism: The Next Decade*, Oxford: Butterworth Heinemann.

# 6 Strategies for development: the role of planning in tourism

Implicit in many perspectives upon sustainable tourism – and indeed, on tourism development in general – is the view that planning has a key role to play in resolving many of the conflicts that such developments may generate. Planning, in its different forms, can be a mechanism for:

- integrating tourism alongside other economic sectors;
- shaping and controlling physical patterns of development;
- conserving scarce or important resources;
- providing frameworks for active promotion and marketing of destinations.

In the absence of planning there are evident risks that tourism development will become unregulated, formless or haphazard, inefficient and likely to lead directly to a range of negative economic, social and environmental impacts.

This chapter attempts three tasks. The first sections aim to explore the basic nature of planning processes and some of the types of planning approach that have been applied to tourism development. Second, the importance of planning tourism is explained, and some of the main strengths and limitations in both conception and implementation of tourism plans are highlighted. Finally, the differences in approach to tourism planning at national, regional and local levels are described and illustrated.

# Planning and planning processes

'Planning' has been defined in various ways, but a common perspective recognises it as an ordered sequence of operations and actions that are designed to realise either a single goal or a set of inter-related goals and objectives. This conceptualisation implies that planning is (or should be) a process:

- for anticipating and ordering change;
- that is forward-looking;
- that seeks optimal solutions to perceived problems;
- that is designed to increase and (ideally) maximise possible developmental benefits, whether they be physical, economic, social or environmental in character;
- that will produce predictable outcomes.

From this broadly based definition, it follows that planning (including planning for tourism) may take on a variety of forms and may be deployed in a great diversity of situations including physical and economic development, service provision, infrastructure improvement, marketing and business operations.

# A general model of the planning process

Although there are a diversity of potential applications for planning, the basic nature of the planning process is remarkably uniform, even allowing for the variation in detail that will reflect the specific applications in which planning is being exercised. Figure 6:1 sets out a general model of the planning process in which the principal elements in devising and implementing a plan are envisaged as a series of key stages.

There are several features of the general planning model to emphasise:

1 There is a progression within the planning process from the general to the specific. The process begins with broad goals and refines these to produce specific policies for implementation.
2 There is an evident circularity in the process by which objectives and the options for realising those objectives are open to review and amendment in the light of either background analysis or the performance of the plan in practice. This links directly to:
3 The dynamic quality of the process. The general model maps out a set of procedures that allow planning to be adaptive to changing

circumstances, a quality that is especially important to tourism planning, where patterns of demand and supply are often volatile. Flexibility should be a key concept for tourism planners.

**Figure 6:1** *General sequence for the production and implementation of a plan*

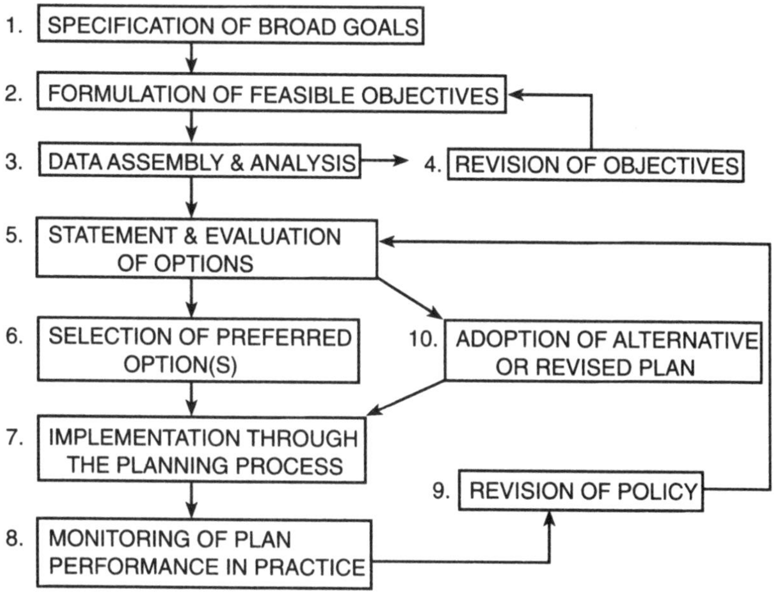

1. SPECIFICATION OF BROAD GOALS
2. FORMULATION OF FEASIBLE OBJECTIVES
3. DATA ASSEMBLY & ANALYSIS → 4. REVISION OF OBJECTIVES
5. STATEMENT & EVALUATION OF OPTIONS
6. SELECTION OF PREFERRED OPTION(S)
10. ADOPTION OF ALTERNATIVE OR REVISED PLAN
7. IMPLEMENTATION THROUGH THE PLANNING PROCESS
9. REVISION OF POLICY
8. MONITORING OF PLAN PERFORMANCE IN PRACTICE

## Types of plan

The general model defines a typical process out of which may be derived many different types of plan or planning approach. Space precludes a detailed discussion of these variations, but to draw some basic points of contrast and comparison, three approaches that will be encountered in the application of planning in tourism are outlined in what follows: master plans, incremental plans and systematic plans.

The master plan approach is arguably the most traditional and also the least suited to the particular requirements of tourism. Master plans centre on the production of a definitive statement that provides a framework for guiding development. The plan defines an end-state (or set of targets) towards which public and/or private agencies are encouraged (or required) to work. Targets are normally expected to be attainable within set time periods – typically a five-year time horizon – and once set in motion, a master plan is normally left to run its course until its time has elapsed. At the end of the plan period, a new master is produced. The master plan approach has the advantage of adopting a comprehensive view of development processes but has also been widely criticised as being too rigid, inflexible and ultimately unrealistic – not least in the guidance of a variable activity such as tourism.

The natural dynamism in tourism (whereby new tourists and new tourism products and destinations tend to redefine patterns more or less continuously) has encouraged some tourism planners to move away from a master plan approach and towards the more adaptable forms of incremental (or continuous) planning. The key difference between incremental plans and master plans is that whereas the master plan is a periodic exercise, incremental planning recognises a need for constant adjustment of development process to reflect changing conditions. So whereas the master plan approach, in defining a blueprint for development, would place an emphasis upon Stages 1 and 2 of the general model (specification of broad goals and objectives), the incremental approach shows a much greater concern for Stages 8–10 (monitoring, revision of policy and objectives, and adoption of revised plans). Since one of the primary objectives of tourism planning is to match levels of demand to supply, this capacity to adjust planning programmes as required is a particular advantage.

One of the recurring themes in the tourism planning literature is the need to plan such a diffuse activity comprehensively and in a manner that integrates the planning of tourism with the other sectors with which it has linkages. Given the breadth of those linkages and the diverse impacts that tourism tends to generate, a planning approach that is comprehensive yet allows for the need for regular readjustment in physical development, service delivery and visitor management is clearly advantageous. Some writers believe such an approach is provided by systems planning.

The systems approach (which originated in the science of cybernetics but is now applied widely in a range of investigative, managerial and planning contexts) is founded on the recognition of interconnections between elements within the system, such that change in one factor will produce consequential and predictable change elsewhere within the system. Thus in order to anticipate (or plan for) change, the structure and workings of the system need to be fully understood and taken into account in any decision-making. In a planning context, systems approaches attempt to draw together four key elements – activity, communications, spaces and time – and map the interdependence between these in producing patterns of development.

The advantages of a systems approach to planning are that it is comprehensive, flexible, integrative and realistic, as well as being amenable to implementation at a range of geographic scales. On the negative side, however, a systems approach requires a great deal of information in order to comprehend how the system actually works (Stage 3 of the general model); it is dependent upon high levels of expertise on the part of the planners and is, therefore, an expensive option to implement. For these reasons it remains the least widely applied of the three methods described, although as planning techniques become more developed, it is an approach that is likely to become more prominent through time.

## Tourism and planning

Planning is important in tourism for a wide range of reasons. First, through the capacity of physical planning processes to control development, it provides a mechanism for a structured provision of tourist facilities and associated infrastructure over quite large geographic areas. This geographic dimension has become a more significant aspect as tourism has developed. Initially, most forms of tourism planning were localised and site-specific, reflecting the rather limited horizons that originally characterised most patterns of tourism. But as the spatial range of tourists has become more extensive as mobility levels have increased, planning systems that are capable of co-ordinating development over regional and even national spaces have become more necessary.

Second, in view of the natural patterns of fragmentation within tourism, any systems that permit co-ordination of activity are likely to become essential to the development of the industry's potential. This fragmentation is mirrored in the many different elements that are required to come together within a tourism plan, including accommodation, attractions, transportation, marketing and a range of human resources (see Figure 6:2), and, given the diverse patterns of ownership and control of these factors in most destinations, a planning system that provides both integration and structure to these disparate elements is clearly of value. Planning systems (when applied in a marketing context) will also enable the promotion and management of tourism places and their products, once they are formed.

**Figure 6:2** *Principal components in a tourism plan*

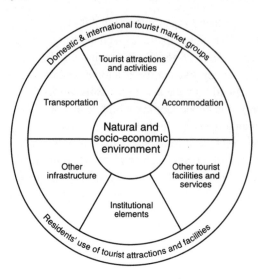

Source: Reprinted from Inskeep, E. *Tourism Planning: An Integrated and Sustainable Development Approach.* Copyright (1991), by permission of John Wiley & Sons Inc.

Third, as noted in the introduction to this chapter, there are clear links between planning and principles of sustainability. Implicit in the concept of sustainable tourism are a range of interventions aimed not only at conserving resources upon which the industry depends, but also at maximising the benefits to local populations that may accrue through proper management of those resources. The commonest form of intervention is via a tourism development or management plan.

Fourth, planning can be a mechanism for the distribution and redistribution of tourism-related investment and economic benefits. This is a particularly important role for planning given that tourism is becoming an industry of global significance but one where activity does not fall evenly across different regions and where the spatial patterns of tourist preference are also prone to variation through time. Planning may assist both the development of new tourist places and, where necessary, the economic realignment of established places that tourists have begun to desert.

Fifth, the integration of tourism into planning systems gives the industry a political significance (since most planning systems are subject to political influence and control) and therefore provides a measure of status and legitimacy for an activity that has not always been taken too seriously as a force for economic and social change.

Lastly, a common goal of planning is to anticipate likely demand patterns and to attempt to match supply to those demands. Furthermore, through the exercise of proper controls over physical development and service delivery, planning will aim to maximise visitor satisfaction. There is now ample evidence from around the world that the unplanned tourist destinations are the ones that are most likely to be associated with negative impacts and low levels of visitor satisfaction, whereas the application of effective planning has often enhanced the tourism product, to the benefit of both host and visitor alike.

This diversity of roles and functions does, however, lead to problems in defining the essential dimensions of tourism planning. In fact, tourism planning, as a concept, is characterised by a range of meanings, applications and uses. It encompasses many activities; it addresses (but does not necessarily blend) physical, social, economic, business and environmental concerns and in so doing involves different groups, agencies and institutions with their own particular agendas. Tourism planning may be exercised by both the public and the private sectors and be subject to varying degrees of legal enforcement. It also works at local, regional, national and (occasionally) at international scales. To talk of 'tourism planning' as if it were a single entity is, therefore, highly misleading, and Table 6:1 attempts to summarise a cross-section of applications that are located within the broad realms of tourism planning.

Apart from ambiguities over what may actually constitute tourism planning, there are further constraints and weaknesses to be taken into account. These include a tendency towards short-termism; organisational deficiencies; and problems of implementation.

The adoption of short-term perspectives is a common characteristic in tourism, and, in the view of some authors, has limited the development of longer-term, strategic planning for tourism. The primacy of short-term responses arises for several reasons. It is a reflection of the natural rhythm of annual cycles within tourism whereby the industry tends to adopt a season-by-season perspective on its performance. But it is also a consequence of the structure of the industry at most destinations and the dominance of small enterprises – a sector that adheres strongly to short-term, tactical views of tourism and is difficult to integrate into wider, longer-term planning frameworks.

Those frameworks may themselves be subject to a range of organisational shortcomings. In many destination areas, the speed with which the need for tourism planning has grown has outstripped the development of organisations, expertise and knowledge to undertake the task. Studies of tourism planning in some of the newer global destinations such as New Zealand and the micro-states of the South Pacific, for example, reveal common problems of inconsistencies in the development of tourism strategies both between and within states and regions; fragmentation and division of responsibility between different public and private agencies; lack of knowledge of patterns of tourism in

**Table 6:1** *Diversity of tourism planning*

| Planning sector | Typical tourism planning concerns/issues |
|---|---|
| Physical (land) | Control over land development by both public and private sectors |
| | Location and design of facilities |
| | Zoning of land uses |
| | Development of tourist transportation systems |
| | Development of public utilities (power, water, etc.) |
| Economic | Shaping spatial and sectoral patterns of investment |
| | Creation of employment |
| | Labour training |
| | Redistribution of wealth |
| | Distribution of subsidies and incentives |
| Social/cultural | Social integregation/segregation of hosts and visitors |
| | Hospitality |
| | Authenticity |
| | Presentation of heritage and culture |
| | Language planning |
| | Maintenance of local custom and practice |
| Environmental | Designation of conservation areas |
| | Protection of flora and fauna |
| | Protection of historic sites/buildings/environments |
| | Regulation of air/water/ground quality |
| | Control over pollution |
| | Assessment of hazards |
| Business and marketing | Formation of business plans and associated products |
| | Promotional strategies |
| | Advertising |
| | Sponsorship |
| | Quality testing and product grading |
| | Provision of tourist information services |

localities; and an absence of planners with specialist knowledge of the industry. Yet even destinations with well-developed planning structures and a good understanding of the tourism markets are not immune from these difficulties. In the UK responsibility for 'planning' tourism falls to a range of agencies (including regional tourist boards, national park authorities and local government planning departments), the last of which rarely contain tourism experts. As a result, the emergence of what some writers have termed an 'implementation gap' – that is, a divergence between what is intended by a tourism plan and what is actually delivered – has been a problem in many localities.

## Tourism planning at national, regional and local levels

The use of geographical scale is a particularly valuable device for drawing out key differences in emphasis and application within tourism planning, and to illustrate the point, the chapter now addresses tourism planning at the national, regional and local levels. However, before we engage with that discussion, three general points are worth noting.

First, although we may distinguish various geographic scales of planning intervention in tourism, these should be seen as interconnected rather than separate spheres of development. In a model framework, such a relationship might be viewed hierarchically: national policies set a broad agenda for development that directly shapes regional-level policies, whilst these in turn form a framework for locally implemented plans. As the scale of intervention diminishes, so the level of detail in planning proposals increases, but the overall aims of planning at each level remain complementary and consistent in direction (Figure 6:3).

In practice, though, neat hierarchical arrangements are rarely found, sometimes because one of the tiers is missing or only partially developed, or, where all tiers are in place, differences in political or institutional attitudes at the different levels may frustrate implementation. In the UK there is no clearly defined regional level of planning so that the regional tourism strategies devised by the tourist boards have been limited in their effect by the absence of legal frameworks for their implementation. In contrast, in New Zealand, concerted attempts to produce regional tourism strategies have been frustrated by the absence of clear policy at the national level. Geographic area will also be a factor, with the absence of a regional tier being especially typical in small nations where regional subdivisions of the national space offer no particular logic or advantage.

Second, in view of the interconnectivity between the different scales of planning, it follows that some areas of concern will form strands that run across all three levels, albeit with varying degrees of emphasis. Economic considerations are one element that may provide a focus of

Figure 6:3  *A model planning hierarchy*

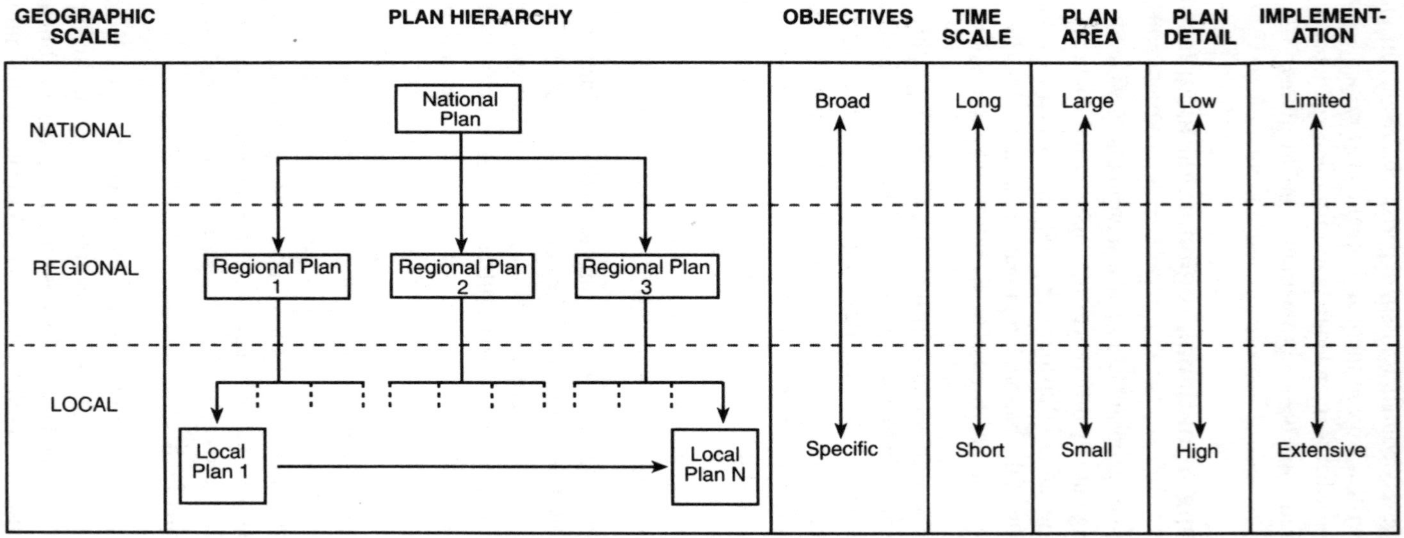

| GEOGRAPHIC SCALE | PLAN HIERARCHY | OBJECTIVES | TIME SCALE | PLAN AREA | PLAN DETAIL | IMPLEMENT-ATION |
|---|---|---|---|---|---|---|
| NATIONAL | National Plan | Broad | Long | Large | Low | Limited |
| REGIONAL | Regional Plan 1   Regional Plan 2   Regional Plan 3 | | | | | |
| LOCAL | Local Plan 1   Local Plan N | Specific | Short | Small | High | Extensive |

interest at all three scales, as are concerns for infrastructure improvements such as transportation and public utilities.

Third, it is inevitable that given the widely differing developmental situations in which tourism planning is applied, there will be marked differences *within* as well as between levels, from place to place. The following discussion should therefore be treated primarily as a generalisation of planning at the three levels, with allowance made for the capacity of individual states, regions or localities to vary substantially from the patterns described.

## Tourism planning at the national level

The significance accorded to national-level planning of tourism varies considerably between destinations but is typically conceptual in character and normally seeks to define primary goals for tourism development and identify policies and broad strategies for their implementation. Within this framework, however, several more specific emphases may emerge, and, in particular, we should note a common concern in national tourism plans with economic issues. This reflects the perceived capacity of international tourism to affect positively a country's balance of payments account and to create employment. Consequently, a growing number of nations, especially in the developing world, have positioned tourism centrally within their national economic development plans.

A second common role for national tourism plans is the designation of tourism development regions. This may be done for any of several reasons: to help structure programmes for the redistribution of wealth and to narrow inter-regional disparities; to create employment in areas where unemployment is an issue; or to channel tourism development into zones that possess appropriate attractions and infrastructure and are therefore considered suitable for tourism. As well as reflecting economic concerns, regional designation may also be guided by environmental factors, in particular a need to protect fragile regions from potentially adverse effects of tourism development.

A third focus of national-level tourism planning is marketing, and this is especially prominent amongst developed destinations that possess the expertise and the resources to form and promote a distinctive set of national tourism products. The strategic planning of British tourism development at the national level is largely absent and the primary role of national agencies such as the British Tourism Authority (BTA) is the marketing of British destinations to domestic and, especially, foreign travellers.

These economic and marketing roles of national level plans are reflected across the globe. Table 6:2 summarises findings from a study of national tourism policies in some forty-nine countries world-wide and places in rank order the eight most important determinants shaping national-level tourism planning in those countries. In addition to emphasising the economic and marketing functions already mentioned, Table 6:2 also draws attention to national issues that occur more selectively. For example, some national tourism plans reflect needs to improve and develop infrastructure, especially transport; others include provision for educational and employment training schemes; whilst a smaller number recognise the potential for tourism to forge international linkages and to maintain positive images of a country within the international community.

However, the approaches to delivering the objectives set out in Table 6:2 vary considerably between nations. Some destinations have adhered to quite rigid programmes of five-year national tourism plans of the kind adopted in countries such as Indonesia, Thailand and Tunisia, whilst others, such as the UK, prefer a more low-key, flexible approach of policy guidance. The physical planning of tourism in England and Wales is only loosely shaped by government Planning Policy Guidance (PPGs), documents which are effectively memoranda to local government planning departments that set out key issues to be addressed and preferred pathways for development, but which allow considerable leeway for local interpretation of the guidance.

**Table 6:2** *Main determinants of national tourism plans and policies in forty-nine countries (in rank order)*

- To generate foreign revenue and assist balance of payments
- To provide employment
- To improve regional and local economies
- To create awareness of the destination/country
- To support environmental conservation
- To contribute to and guide infrastructure development
- To promote international contact and goodwill

Source: Baum (1994).

Institutional contexts of national tourism planning are also variable. In the study of national tourism planning referred to in Table 6:2, only half of the countries surveyed had established a government department with sole responsibility for national tourism planning and nearly 15 per cent apparently had no governmental-level interests in the sector at all. Elsewhere, tourism was accorded only secondary interest, and this is often reflected in movement (or division) of tourism planning briefs between government departments. In the United Kingdom in the ten years between 1983 and 1992, tourism development was first the responsibility of the Department of Trade and Industry, then the Department of Employment. It is now a part of the newly formed Department of Culture, Media and Sport (originally called the Department of National Heritage). This rather uncertain status reflects the secondary position that tourism holds in many national planning frameworks and is a weakness that, in the longer term, may well need to be addressed.

To illustrate a rather more structured approach to tourism planning at the national level, Box 6:1 describes the planning of tourism in one emerging destination – Tunisia.

## Tourism planning at the regional level

In comparison with national forms of tourism planning, regional tourism plans are usually distinguished by a marked increase in the level of detail and a sharper focus upon particular developmental issues. National plans tend to be broad statements of intent, but at the regional level the implications of those intents can be mapped far more precisely and planning can reflect specific requirements. Since the implications of development proposals for individual localities also become more apparent, some degree of public interest or participation within the tourism planning process may also be evident.

Several themes are likely to be carried through from the national to regional levels; in particular:

- concerns for the impact of tourism upon regional economies and employment patterns;
- development of infrastructure, including transport systems to assist in the circulation of visitors within the region, as well as provision of public utilities such as power and water supplies, both of which are frequently organised at regional levels;

## Box 6:1

### *National tourism planning in Tunisia*

Tunisian development has been guided for many years by a succession of five-year National Development Plans with the most recent – the eighth – due to terminate at the end of 1997. The development of tourism has been an integral and increasingly significant element within these plans, as the role of tourism within the Tunisian economy has become more central and as the scale of the industry has increased from a mere 50,000 foreign visitors in 1962 to more than 3 million visitors in 1990.

The Tunisian National Development Plans conform largely to the master plan concept insofar as a major element in the planning approach is the designation of targets for growth and investment. For example, the Seventh National Development Plan (1986–91) set the following targets for tourism:

- bed spaces to increase by 19 per cent to 118,000;
- bed occupancy to increase by 42 per cent to 18 million bed-nights;
- direct employment to increase by 13 per cent to 46,000;
- cumulative investment to increase by 72 per cent to 1,243 million Tunisian dinars (approximately £777 million);
- annual tourism receipts to increase by 104 per cent to TD 797 million (£498 million).

Whilst much of the actual provision to support these targets was expected to come from private-sector investments, direct government intervention contributed significantly to the realisation of the plan objectives via a number of pathways. These included:

- investment in infrastructure (especially transportation);
- promotion and marketing (which has been particularly important following recession in the 1980s and a temporary slump following the Gulf War);
- training programmes (which had previously established training schools in all the regions and, under the Seventh Plan, added a new hotel school at Monastir);
- regional initiatives aimed at diversification of the tourism product and development of new tourism areas.

For tourism planning and investment purposes, Tunisia is divided into seven regions, each of which has its own growth target under the National Plan (Figure 6:4). To date, most tourism development has centred on the north-eastern coastline around Tunis and the Bay of Hammamet, but under the Seventh National Plan, several new tourism areas were announced. These included a new integrated resort at Port-el-Kantaoui with over 13,000 bed spaces, a marina, restaurants and a range of sports facilities, together with smaller schemes planned for Hergla and Gamarth. However, more significantly, projects at Tabarka and Bizerte on the relatively undeveloped northern Tunisian coast were also announced, along with proposals for new tourist access to the arid interior in southern Tunisia. At Tabarka a new international airport provides an additional gateway to the area whilst the Gafsa–Tozeur region in the south has seen 15 million Tunisian dinars (£9.3 million) invested in the creation of a new integrated tourist route (*Chaîne hotelière caravaneserail*) which links the coast with desert and mountain oases at Tamerza. This follows old Arab trading routes and uses accommodation in modern versions of the traditional caravaneserai – a form of hotel or staging post. This particular initiative is aimed not only at promoting new destinations within Tunisia but also at appealing to a different clientele: groups that are looking for more than conventional resort-based beach holidays.

Source: Gant and Smith (1992).

**Figure 6:4  *Tourism development in Tunisia***

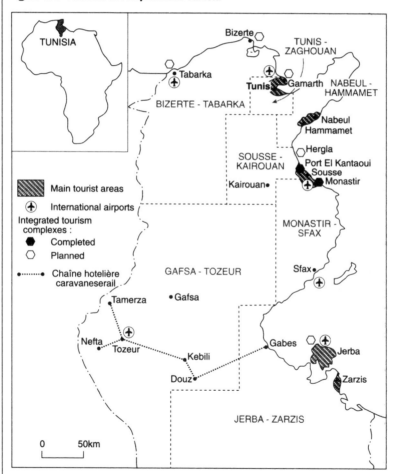

- further spatial structuring in which tourism localities within regions are identified;
- regional-level marketing and promotion, especially where the region possesses a particular identity and/or set of tourism products.

However, there will also be distinctive features of a regional tourism plan that may not be found at a national level. First, regional plans commonly show greater levels of concern over environmental impacts. Except in the case of small nations, the uneven spatial patterns that are associated with tourism mean that environmental impacts are seldom felt at a national level but are manifest within regions and localities. The tendency of environmental effects to spread through natural systems and across wider geographic spaces (see Chapter 5) also means that a regional scale of planning is often the appropriate level for intervention with planned solutions to such impacts. (The attempt at producing sustainable planned development of Australia's Great Barrier Reef outlined in Box 5:3 provides an example.)

Second, regional plans will often contain detailed consideration of the type and location of visitor attractions, together with supporting services such as accommodation. Such matters are rarely articulated in detail within a national tourism plan, but in regions the reduction in geographic scale makes it easier to define locations that will support tourism, establish how far existing capacities match expected demand, and thus plan developments of new attractions and services that are required to meet identified deficiencies.

Third, regional plans may reflect needs associated with the management of visitors. Distinctions between management and planning in tourism are often blurred, but unless the nation is small, the regional level is normally the first point at which tourist management issues begin to emerge clearly, albeit still at a macro scale. Regional zoning strategies aimed at either concentration or dispersal of visitors, the planning of tourist information services, designation of tourist routes and strategic placement of key attractions may all form part of regional tourism management strategies.

Box 6:2 contains an example of a regional tourism strategy that combines elements of both physical planning and the development of tourism management in the South West of England.

## Box 6:2

### *Regional tourism planning in South West England*

Although the regional tourism strategies produced by tourist boards in England and Wales have no legally enforceable status, they provide important frameworks for co-ordinating public- and private-sector development of tourism and associated visitor management. The South West of England (Figure 6:5) is the major tourism region in the UK, and, with around 14 million staying visitors annually, receives more than twice as many tourists as the next most popular region. But it is also a tourism region beset by problems of congestion and poor accessibility, as well as a legacy of nineteenth-century seaside resorts in which decline in the face of foreign competition and reducing levels of investment has become a major issue.

The stated objectives of the 1992–6 strategy are:

- to identify key issues facing tourism in the region;
- to establish a framework in which actions of public and private agencies can be planned and co-ordinated;
- to raise the profile of tourism to mobilise support and allocation of resources.

The production of the strategy follows a classical pattern in which the context of regional tourism was analysed in detail to identify sets of objectives and to establish priorities for development. The contextual analysis gives consideration to national strategies of the English Tourist Board and also emphasises links between tourism planning and the planning objectives as set down in county structure plans and the plans of the two national parks within the region. The strategy was also informed by analyses of existing tourism resources (accommodation, visitor attractions, natural resources, etc.); market profiles of visitors; analysis of trends in the main market segments (long-stay holidays, short breaks, business tourism, etc.); and the outlook for tourism in the region in the light of those trends and when set alongside an analysis of the strengths and weaknesses in the regional tourism product.

The resulting strategy centred upon three themes:

- spreading tourism benefits both spatially into new areas and temporally by extending the season, which, at present, is highly focused into the traditional summer months and upon conventional seaside tourism. The strategy emphasises the need for new products and new tourism places to reinvigorate the industry across the region;
- conserving and enhancing the environmental and cultural resources that form a basis to tourism in the region;
- enhancing customer satisfaction by emphasis upon quality and value for money, whilst simultaneously addressing problems such as traffic congestion, which the strategy identifies as a major negative element in visitors' perceptions of the region.

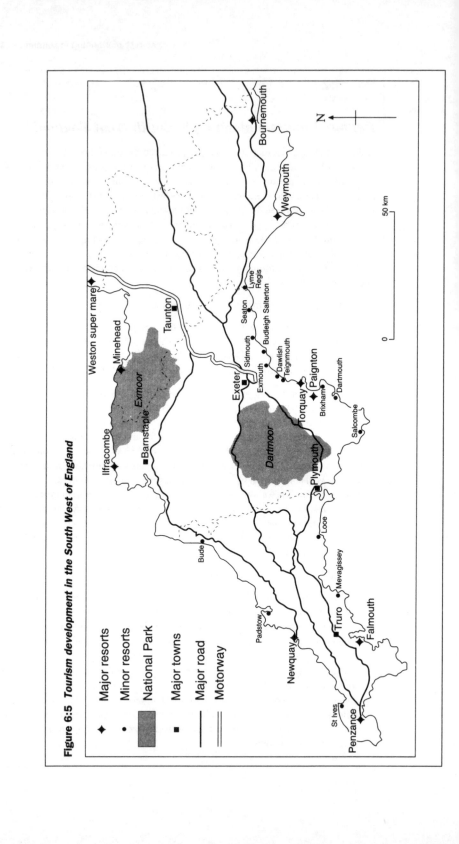

Figure 6:5 *Tourism development in the South West of England*

**Key**

| | |
|---|---|
| ◆ | Major resorts |
| ● | Minor resorts |
| ▨ | National Park |
| ■ | Major towns |
| — | Major road |
| ═ | Motorway |

0                                    50 km

These broad goals are encapsulated in a detailed discussion of constraints, opportunities and priorities across nine broad sectors of the industry: resorts and urban places; the countryside; accommodation; attractions; sport and the arts; transport and access; marketing; information; and training and support services. In each section the subject is related to the strategic objectives, and a series of actions judged capable of meeting those objectives is set out.

For example, the strategy for the resorts (which is set against the background of their sustained decline as patterns of tourism have shifted away from domestic seaside holidays) places priority upon tackling their special needs and strengthening their appeal. To meet these objectives, each resort is encouraged to:

- explore specific target markets, particularly in business and special-interest tourism;
- upgrade accommodation and other visitor services such as catering;
- foster events and attractions that will encourage low-season visits;
- develop links with attractions in rural hinterlands;
- improve visitor management.

To reinforce these ideas, the strategy advocates paying greater attention to environmental concerns, including the cleanliness and management of beaches and inshore waters as well as the quality of waterfront, quays and docks in coastal towns. The latter are seen as having clear potential for restoration and conversion to meet tourist functions. Zones of land with particularly attractive buildings, land patterns or landscape are proposed for designation as 'core areas' in which both public and private investment should be concentrated with the aim of enhancing the quality and character of resorts.

The future prosperity of resorts is also seen as resting upon effective visitor management with particular needs for improvements in the capacity and location of vehicle parking; the visibility and usability of tourist information points; pedestrian signposting; and cleanliness of street areas and public toilets.

As with many tourism strategies, however, success depends upon effective implementation, and whilst the regional tourist board possesses some funds to support selected projects (especially in promotional work and activity such as quality appraisal), full implementation is dependent upon the establishment of effective partnerships between local authorities, private enterprise and the tourist board. The ability to form such partnerships is understandably variable.

Source: West Country Tourist Board (1991).

## Tourism planning at the local level

Local-level planning of tourism is a highly variable activity, reflecting the diversity of local situations in which tourism is developed. Yet at the same time, this level of application is also the easiest at which to identify a core of common planning concerns.

Most forms of local tourism plan are primarily focused upon the physical organisation of tourism resources (accommodation, local transport, catering and local attractions), the control of physical development (such as hotel construction) and the exercise of local visitor management. Local plans are typically short term and regulatory in nature (rather than being longer-term, strategic statements) with particular concerns for reducing development conflicts and harmonising activities that use the same spaces and/or resources. Local plans will show some similarities to regional-level plans in their attention to the logistics of provision of supporting infrastructure – power supplies, water and sanitation, local accessibility, etc. – but will be distinctly more detailed in their approach. Unlike regional plans, however, local planning of tourism will pay much greater levels of attention to the physical design and layout of developments – something that is rarely encountered at the larger geographic scales of intervention.

Local planning is often seen as the most effective level for the implementation of physical land use plans and associated tasks such as the spatial zoning of activity and developments. This is for two reasons. First, it is the planning level at which there is most likely to be a legally enforceable system of planning control. In England and Wales, for instance, although the county structure plans map the broad planning strategies at a regional or sub-regional scale, implementation is mainly via the development control process, which is operated through the medium of local plans and local decisions on development. Second, in most cases the appropriateness of a proposed development is most effectively judged in a local context, since this is the level at which impacts are to be most clearly felt. For this reason, it is also the level at which questions of public reaction to development are best considered, as the implications of proposed developments become prominent and measurable. As a result, local plans may take more account of socio-cultural impacts than will national- or regional-level strategies, although, given the nature of social effects of tourism (see Chapter 7), this dimension is difficult to encompass in planning processes, however they are structured.

Although controlling development is an important and distinctive function of local plans, they may also reflect issues that are addressed at regional and national levels, especially economic and environmental effects of tourism. Economic impacts are likely to be considered in terms of scope for local employment, new-firm formation and potential multiplier effects of tourism incomes. Environmental and conservation issues will also be addressed, especially since the existence of legal controls and the increased use in many local planning procedures of EIA (see Chapter 5) reinforce the capacity of local planning systems to protect conservation areas and fragile environments from potentially harmful physical developments.

Given the diversity of local situations, local plans can take on many nuances in character and purpose, including a strong role in visitor management. However, one of the commonest applications of local planning in tourism is in relation to resort development. This may be applied both to programmes to redevelop existing resorts that may be in decline, and to the construction of the new generations of integrated resorts that characterise many of the emerging tourism destinations. Box 6:3 therefore aims to illustrate one application of the local planning approach by outlining an example of a planned, integrated resort in Hawaii. The example also illustrates a form of comprehensive planning that embodies some elements of the systematic approach discussed earlier in this chapter, albeit at a very localised scale.

---

## Box 6:3

### *Local tourism planning: Mauna Lani, Hawaii*

The Mauna Lani resort is an example of a modern, multi-functional resort which was carefully planned from inception to offer high-quality accommodation and a range of recreational facilities, set within an attractive environment that included significant natural and non-natural resources. It is located on the north-west coast of the main island of Hawaii and is one of several local resort developments within an area designated in the regional plan as being suited to this form of provision. To reinforce the viability of this new tourism area, the state government invested substantially in new public infrastructure, including new road and air access together with a new water distribution system, as part of its regional plan.

Initial development by a private company of a 778-acre (315-ha) site at Mauna Lani was given local approval in 1980, with a plan for 3,000 hotel rooms and a further 3,182 bed spaces in bungalows and condominiums. The plan also allowed for an 18-hole golf course, on-site shopping and entertainment, with extensive parkland and trails. A second phase of development (approved in the late 1980s) saw the site expand to 1,432 acres (580 ha), primarily to allow the provision of a second golf course and additional amenity land (Figure 6:6). In order to reduce residential density, the bed space limits in the expanded resort remained the same

**Figure 6:6  *The Mauna Lani resort, Hawaii***

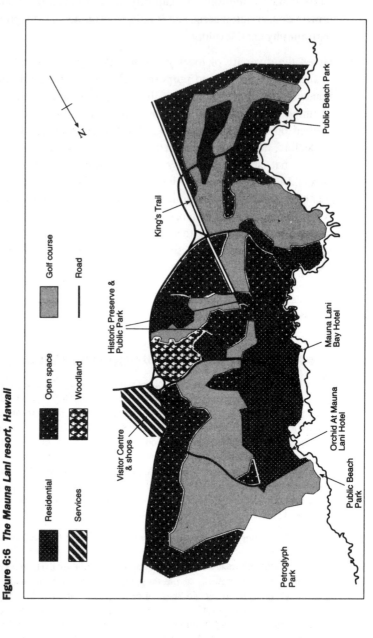

Source: Adapted from Inskeep (1991).

as approved in the initial plan, and so that the enlarged resort did not infringe traditional local rights of access, two areas of coastline which would retain open public access were agreed.

Two features of the plan are worth noting. First, attention to design and layout was central to the conception of the resort as a quality leisure environment. This is shown by the manner in which:

- Open spaces and the golf courses are used to create an attractive environment with varied combinations of buildings and amenity spaces, together with views, especially of the sea. Placement of buildings and routeways was carefully designed to take advantage of the natural form of the terrain and to set out corridors of amenity land through the resort.
- The main commercial facilities and reception areas are accessible but not centrally located, whilst service areas (offices, stores and the resort's own sewage treatment works) are discretely placed to one side of the development. This was a conscious decision aimed at encouraging a relaxing and leisurely ambience in the principal areas of the resort.
- The roadways, whilst providing access, do not facilitate through traffic, nor do they encourage routine use of vehicles inside the resort.

The second important aspect of the plan was the manner in which the developers were required to retain several important environmental features on the site, in particular the shoreline, a number of historic landscapes with ancient fishponds, and old walkways, such as the 'King's Trail'. As part of the planning agreement, the developers also undertook to maintain and develop suitable visitor facilities in the adjacent Puako Archaeological Petroglyph Park (a conserved area of volcanic rock formations and rock carvings).

This example demonstrates a number of features of local-level tourism planning:

- the importance of careful design and location of facilities;
- the capacity of local planning to protect an environment with detailed and site-specific adaptation of facilities to a location and incorporation of natural and non-natural features into overall design concepts;
- the ability of authorities to draw concessions (or what is sometimes termed 'planning gain') from developers in return for permission to develop – in this case, the agreement of the developer to manage the Petroglyph Park.

Source: Inskeep (1991).

These examples and case studies are intended to illustrate the range of applications of planning in tourism development. It is important to realise that the preceding text provides only an outline review of a truly extensive topic. Readers with a desire to probe the subject more deeply are therefore encouraged to explore the literature cited below and in the overall Bibliography.

# Summary

By focusing on the role of planning in shaping physical developments, the chapter highlights those aspects of tourism planning in which geographical perspectives are most useful in delivering an understanding of processes of change, although clearly, this is not the only way in which tourism is influenced by planning processes. The chapter also illustrates that tourism planning is an overtly geographic phenomenon, varying not only through time but, more significantly, across space. Planning at national, regional and local scales is now widely encountered, and whilst there are common themes and issues that link the different scales of intervention, there are also distinctive dimensions that typify planning for tourism at these different spatial levels.

# Discussion questions

1 What are the main differences in approach and emphasis that are likely to occur in tourism planning at national, regional and local levels?
2 To what extent is the effectiveness of tourism planning inhibited by the nature of tourism itself?
3 Tourism planning often reveals some common deficiencies. What are they and why do they arise?
4 How can planning help tourism become sustainable?

# Further reading

Full and comprehensive consideration of tourism planning in its many guises may be found in:
Gunn, C.A. (1988) *Tourism Planning*, New York: Taylor and Francis.
Inskeep, E. (1991) *Tourism Planning: An Integrated and Sustainable Development Approach*, New York: Van Nostrand Reinhold.

Useful discussions of tourism planning written from a geographical perspective are provided in:
Murphy, P.E. (1985) *Tourism: A Community Approach*, London: Routledge: 153–176.
Pearce, D.G. (1989) *Tourist Development*, Harlow: Longman: 244–279.

Convenient reviews of national and regional tourism strategies in a range of destinations are contained in:
Baum, T. (1994) 'The development and implementation of national tourism policies', *Tourism Management*, Vol. 15 No. 3: 185–192.

World Tourism Organization (1994) *National and Regional Tourism Planning*, London: Routledge.

An excellent discussion of the links between tourism planning and sustainability is provided by:
Hunter, G. and Green, G. (1995) *Tourism and the Environment: A Sustainable Relationship?*, London: Routledge: 93–121.

Recent case studies of tourism planning in practice are to be found in:
Briguglio, L., Butler, R.W., Harrison, D. and Filho, W.L. (eds) (1996) *Sustainable Tourism in Islands and Small States: Case Studies*, London: Pinter.
Cooper, C.P. (1995) 'Strategic planning for sustainable tourism: the case of the offshore islands of the UK', *Journal of Sustainable Tourism*, Vol. 3 No. 4: 191–207.
Page, S.J. and Thorn, K.J. (1997) 'Towards sustainable tourism planning in New Zealand: public sector planning responses', *Journal of Sustainable Tourism*, Vol. 5 No. 1: 59–75.

# 7 Cultures and communities: the socio-cultural relationships between hosts and visitors

In this chapter we explore some of the links between tourism, society and culture. The social dimensions to tourism and the attraction of different cultures as a motivation for travel are well established. For example, although the first sea bathing resorts were initially formed as health spas, they were soon transformed into fashionable *social* environments, and, as we have seen in Chapter 3, the Grand Tourists of the seventeenth and eighteenth centuries – precursors of the modern international tourist – shaped their itineraries around the major cultural sites of Europe, particularly Italy. Although the significance of the cultural novelty of different destinations has fluctuated as a motive for travel – especially as mass forms of seaside tourism emerged during the last quarter of the nineteenth and the first half of the twentieth centuries – it has seldom disappeared. Indeed, within the increasingly globalised and uniform contemporary lifestyles of the developed nations that still generate most of the world's international tourists, the appeal of local foreign cultures, with their distinctive traditions, dress, languages, handicrafts, food, music, art and architecture, has never been stronger. Culture and the societies that create culture have become central objects of the tourist gaze.

However, the relationships between tourism and culture do not simply revolve around the role of culture as an object of tourist attention, but also embrace a wide range of socio-cultural impacts that contact between hosts and visitors may promote. As with the other areas of tourism impact that we have considered, the social consequences of tourism development span the spectrum from positive to negative and, from a geographic perspective, vary significantly from place to place.

Our understanding of the relationships between tourism, societies and cultures is, though, incomplete and uncertain – hampered by both a limited conceptual basis and inconclusive or conflicting empirical studies. Several factors may be held to account for this situation:

1 Uncertainty arises especially from the complexity of processes of socio-cultural change and the near impossibility of filtering the specific effects of tourism from the general influence of other powerful agents of change, such as globalised television and the media.
2 Socio-cultural impacts have received relatively modest attention, partly because most social and cultural beliefs or practices are much less amenable to direct observation and the conventional forms of measurement through survey-based enquiry of the kind that is so popular in the analysis of tourism.
3 For similar reasons, social concerns arising from tourism are often poorly accommodated in planning processes where primary interests centre upon controlling physical development, encouraging economic growth and, more recently, promoting sustainable environments.
4 Neglect is also a consequence of a tacit set of assumptions which still prevail in many tourism areas – namely that social impacts are slight, that communities will adapt and that people will learn to tolerate the socio-cultural changes that tourism brings to their lives as a price worth paying to realise the economic benefits that the industry can create. There are, though, a growing number of case studies that challenge this view and which suggest that for many locations, the socio-cultural impacts of tourism are neither trivial nor unavoidable.

Some form of socio-cultural impact is an inevitable part of the host–visitor relationship because tourism brings together regions and societies that are normally characterised by varying degrees of difference. The visitors will tend to originate in a developed, urbanised and industrialised society and will carry with them the beliefs, values and expectations that such societies promulgate. But as the spatial range over which tourists roam is extended (and given the predisposition of many tourists to seek out places that are different), so the likelihood increases that encounters between hosts and visitors will bring together opposing tendencies and experiences: development and underdevelopment;

pre-industrial, industrial and post-industrial; traditional and
(post)modern; urban and rural; affluence and poverty; etc. It has also
been observed that such encounters are often unequal or unbalanced in
character, not just in terms of material inequalities but in context too: the
visitor at leisure and probably enjoying novel situations, whilst the host
pursues the familiar routine of work.

Much of the academic literature on the socio-cultural impacts of tourism
tends to emphasise negative perspectives. This can, however, be
misleading and, as we shall see in the second half of this chapter, tourism
development can foster positive effects too. Popular views of tourism as
the 'destroyer' of societies and their cultures are too simplistic. Tourism
is not a monolithic force, nor does it stand apart from wider processes of
development and change. It is both a cause and a consequence of socio-
cultural development, and since it comprises a diversity of participants,
agencies and institutions with differing motives and goals, its effects are
diverse and often unpredictable. This leads to considerable spatial and
temporal variation in the nature of relationships between tourism, society
and culture and the effects that it creates. However, what is also clear is
that in general, the presence of the visitor changes the object of his or her
attention. This is a paradox that is common to several dimensions of
modern tourism but it underpins many of the concerns that have been
voiced over the socio-cultural impacts of tourism.

## Tourism, society and culture

## Theoretical perspectives

What are the mechanisms through which tourism affects cultures and
societies? A number of theories and concepts have been advanced to
attempt to explain how host–visitor contacts advance socio-cultural
change, but two – the demonstration effect and processes of acculturation
– have proven particularly popular.

The demonstration effect is dependent upon the existence of visible
differences between visitors and hosts as this theory suggests that
changes in the hosts' attitudes, values or behaviour patterns may be
brought about simply through their *observing* the tourist. It is argued that
by observing the behaviours and superior material possessions of
tourists, local people may be encouraged to imitate actions and aspire to

ownership of particular sets of goods – clothing, for example, that they see in the possession of the visitors and to which they are attracted. In some cases, the demonstration effect can have positive outcomes, especially where it encourages people to adapt towards more amenable or productive patterns of behaviour and where it encourages a community to work towards things that they may lack. But more typically, the demonstration effect has been characterised as a disruptive influence, displaying a pattern of lifestyle and associated material ownership that is likely to remain inaccessible to local people for the foreseeable future. This may promote resentment and frustration or, in cases where visitor codes and lifestyles are partially adopted by locals, lead to conflicts with prevailing patterns, customs and beliefs. Young people are particularly susceptible to the demonstration effect, and hence tourism has occasionally been blamed for creating new social divisions between community elders and the young in host societies, or the encouragement of age-selective migration, with younger, better-educated people moving away in search of the improved lifestyles that the demonstration effect outwardly displays. The migrant, of course, may well benefit from such a move but the social effects upon the community that is losing its younger members will be broadly detrimental.

The demonstration effect, with its emphasis upon detached forms of influence, is particularly attractive in explaining tourism impacts where contacts between host and visitor are typically superficial and transitory. However, where links between hosts and visitors are more fully developed, acculturation theory offers an alternative perspective. Acculturation theory states that when two cultures come into contact for any length of time, an exchange of ideas and products will take place that will, through time, produce varying levels of convergence between the cultures; that is, they become more similar. The process of exchange will not, however, be a balanced one since a stronger culture will dominate a weaker one and exert a more powerful effect over the form of any new socio-cultural patterns that may emerge. (Interestingly, 'strength' of culture is not necessarily a reflection of cultural distinctiveness or integrity. The USA, for instance, has one of the most pervasive and powerful cultural influences that is spread by tourism, yet American cultural strength is more a reflection of population size, economic power and growing domination of global media than a particularly well-defined cultural identity, US society being strongly multicultural.)

As with the demonstration effect, processes of acculturation are most easily envisaged in relationships between the developed and the developing world, but such patterns may also be found within developed states. Many European nations contain marginal or peripheral regions that are attractive to tourists and which also contain distinctive cultures. Within the UK this is the case for large parts of Wales, where a strengthening local resistance to changes due to acculturation (such as the erosion of the Welsh language) that have in part been associated with tourism has been noted. Processes of acculturation should therefore be expected to operate in a range of spatial contexts.

## Factors promoting variation in tourism impacts

Spatial variation in the nature and consequences of host–visitor encounters will occur for a number of reasons, but we may identify a set of key variables that will help to explain differences in effect. These are the nature of the encounter and visitor type; the nature of the location; spatial proximity of host and guest and levels of involvement in tourism; cultural similarity; and the stage of development.

### Nature of the encounter and visitor type

It has been suggested that visitors and hosts encounter one another in three basic situations:

- when tourists make purchases of goods and services from local people in shops, bars, hotels or restaurants;
- when tourists and hosts share the same facilities, such as local beaches and entertainment;
- when they meet purposely to converse and to exchange ideas, experiences or information.

The extent and the nature of socio-cultural impacts will clearly be influenced by whichever of these forms of contact prevails, but their incidence will also tend to reflect the type of visitor (see Chapter 1) and the periodicity and duration of their visits. When tourism is centred upon mass markets, contacts are most likely to be in either (or both) of the first two categories, but because of the limited seasons associated with many package holidays and the short duration of individual trips, the contacts are typically casual and brief. However, although contact may be limited, the scale of mass forms of tourism is still quite capable of producing a range of problems and changes through the demonstration effect or acculturation. Purposeful engagement between the two groups is much rarer in modern tourism and is more typically a feature of independent traveller and explorer types of tourist. Since they are less numerous, these tourists are generally held to have lesser impacts upon local societies and cultures, although, strictly speaking, any form of contact is likely to produce some degree of social and cultural change, and if 'explorer' types of tourist spend extended periods in a host community, scope for cultural interchange will be significantly increased.

## Nature of the location

Geographic elements are also important, both in the nature of destinations and in the effect of spatial proximity between hosts and visitors. The tolerance of tourism by local communities will be affected by the capacity of a locality to absorb tourism and, more simply, the degree to which tourists form identifiable groups and/or create visible problems. In metropolitan centres such as London or Paris, thousands of tourists may be accommodated with few discernible impacts because urban infrastructures are designed to cope with heavy use and, in many situations, the tourist simply blends with the crowds. In contrast, small rural communities that are not adapted to handling crowds may struggle to cope with more than a few hundred visitors, and because those visitors are far more conspicuous, scope for induced change through demonstration effects and acculturation will be enhanced.

## Spatial proximity and levels of involvement

The nature and the intensity of impacts will be influenced by spatial and sectoral proximity of hosts and visitors. As we have seen, tourism development has a marked tendency to spatial concentration at favoured locations, so patterns of development are uneven. Whilst we would expect some diffusion of impacts from centres of tourism into surrounding areas (for example, through the employment of people who travel daily to work in tourism from a hinterland), the capacity of tourism to affect host societies and cultures will decline as distance from the tourist centres increases. Even within tourism areas, some locations remain untouched by tourists, and their routine movements and certain forms of development – especially enclaves – will purposely segregate hosts and visitors, thereby minimising social and cultural impacts. For similar reasons, different sectors within a local community will react variably to the presence of tourists. Business sectors and government are more likely to adopt favourable views of tourism owing to the economic benefits that the industry can bring, whereas ordinary local residents who do not benefit directly from the tourism economy but whose lives are affected by the noise, overcrowding, congestion and over-use of facilities that tourism often creates will tend to form negative views. Thus attitudes and behavioural responses towards tourism are normally differentiated by the direct or indirect ways in which the various groups within communities experience tourism.

## Cultural similarity

However, perhaps the most important factors in shaping socio-cultural impacts are the levels of cultural similarity or dissimilarity and the stage of tourism development that has been attained. Cultural 'distance' (which often tallies closely with spatial distance) between visitor and host will be crucial in determining the level of impact that is likely to be felt. The maximum social impacts tend to occur when a host community is relatively small, unsophisticated and isolated, and where affluence levels are markedly different. When hosts and visitors have similar levels of socio-economic and technological development, socio-cultural differences will tend to be less pronounced and tourism impacts upon society and cultures are reduced in consequence. Although international tourism does bring differing groups together, in many locations tourism also brings together culturally *similar* people. In North America, for example, interchange between Canadian and American tourists, whose lifestyles have much in common, produces comparatively few socio-cultural repercussions (although impacts upon Native American communities in the USA and Canada may be more pronounced). Even in the rapidly expanding markets of South-East Asia, a region in which cultural impacts might be expected to be an issue, over 75 per cent of international visitors originate within the region. Thus, although there are important differences between the major ethnic groups in this area, there remains a sufficient breadth of shared socio-cultural experiences to produce fewer impacts than might have been anticipated.

However, simple views of host–visitor relationships as centring upon a basic dichotomy of just two cultural forms – the host and the visitor – have been challenged by a number of authors. First, it should be remembered that tourist flows to many destinations will be composed of tourists from a variety of sources with differing cultural backgrounds and contrasting levels of cultural difference. The United Kingdom, for example, receives significant numbers of tourists from Europe, North America and Japan – some of whom are socially and culturally closer to the British than are others.

Second, it is a mistake to assume that destinations are themselves culturally and socially homogeneous. The United Kingdom, once again, is a case in point, with strong regional cultures and traditions, reinforced in Wales and, to a much lesser extent, in north-west Scotland by linguistic contrasts.

Third, and most importantly, however, we should recognise that the behaviour patterns of the visitors are often a diversion from their socio-cultural norms and do not, therefore, accurately represent the host societies from which they originate. As was noted in Chapter 1 (Table 1:1), tourist behaviours often display a conscious departure from normal patterns, with conspicuous increases in levels of expenditure and consumption, together with behavioural inversions that see tourists resorting to activities that might be on the margins of social acceptability at home – for example, drinking and over-eating, gambling, atypical dress codes, nudity or semi-nudity, etc. In other words, there exists within the visitors' normal culture a 'tourist culture' – a sub-set of behavioural patterns and values that tend to emerge only when the visitors are travelling but which, when viewed by local people in receiving areas, projects a false and misleading image of the visitors and the societies they represent.

These ideas are illustrated diagrammatically in Figure 7.1. Each box represents a culture with the tourist culture nesting within, but also extending outside, the normal visitor's culture, representing the tendency to atypical forms of behaviour. The greater the extent of overlap between the cultures, the greater the

**Figure 7:1  *Cultural distance and the socio-cultural impact of tourism***

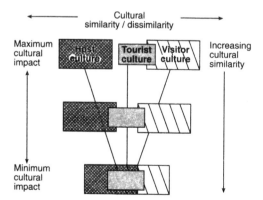

socio-cultural similarities and the fewer the resulting tourism impacts. Conversely, the less overlap that exists between the three, the greater the cultural distance between host and visitor and, consequently, the greater the chance of tourism producing socio-cultural changes.

## Stage of development

In Chapter 2, Butler's model of resort development was used to illustrate how tourism places evolve through time. An important theme that is implicit in Butler's conception is that tourism impacts will also evolve through time, the natural tendency being for the scale of impact to grow as the destination progresses from exploratory stages (where impacts are slight) to the stages of saturation (in which impacts will be significant).

A useful application of the idea of evolution to the cultural and social impacts of tourism has been provided by Doxey's 'Irridex' – a contraction of 'irritation index' – which attempts to show how attitudes to tourism in a host area might change as the industry develops (Figure 7:2). Initially the tourists are welcomed, both as a novelty and also because of the scope for creating economic prosperity. As developments become more structured and commercialised, local interest in the visitors becomes sectionalised (i.e. some local people become involved with the

**Figure 7:2  Doxey's 'Irridex'**

'IRRIDEX'

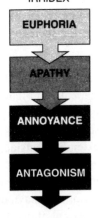

Initial phase of development, visitors and investors welcome, little planning or control mechanism.

Visitors taken for granted, contacts between residents and outsiders more formal (commercial), planning concerned mostly with marketing.

Saturation points approached, residents have misgivings about tourist industry, policy-makers attempt solutions via increasing infrastructure rather than limiting growth.

Irritations openly expressed, visitors seen as cause of all the problems, planning now remedial but promotion increased to offset deteriorating reputation of destination.

Source: Murphy (1985).

tourists, others do not) and signs of apathy emerge, especially amongst the uncommitted. If growth continues, physical problems of congestion and spiralling development sow seeds of annoyance on the part of local people, whose lives are now increasingly affected and inconvenienced by tourism. In the final stage of Doxey's model, annoyance has turned to open antagonism and hostility towards the tourists, who are now blamed, fairly or unfairly, for perceived detrimental changes to local lifestyles and society.

Although it maps a pathway that will certainly be encountered in some tourism destinations, Doxey's model has drawn a number of criticisms. The most significant are that the concept is essentially a negative reading that permits little recognition of positive benefits, whilst the unidirectional quality of the model suggests only an inevitable sequence of decline in the host–visitor relationship. In practice, such relationships are rather more complex and prone to greater variation than the model allows. As we have seen, the attitudes of people who are directly involved with and benefiting from the industry will differ from the attitudes of those who are not, and, in the normal course of events, people will move into and out of tourism as the industry develops and changes. Situations surrounding tourism development may improve and the negative effects that may encourage antagonistic attitudes can be offset by effective planning. Differing societies and cultures will also be more or less adaptable, thereby delaying or even offsetting altogether the latter stages of Doxey's model. The 'Irridex' should not therefore be taken as a definitive description of host–visitor relationships, although it does have a value in highlighting the potential for worsening relationships between tourists and local populations as the industry expands.

## Effects of tourism upon host communities

Empirical studies of the socio-cultural effects of tourism have highlighted a diversity of possible effects, and these are broadly summarised in Table 7:1. Very full discussions of these issues may be found within the general tourism literature, so for the purposes of this book it is proposed to examine and illustrate a cross-section of impacts which, for convenience, will be grouped under three broad headings: authenticity and commodification; moral drift and changing social values; and new social structures and empowerment.

**Table 7:1** *Primary positive and negative impacts of tourism upon host society and culture*

*Main positive impacts*
- Increased knowledge and understanding of host societies and cultures
- Promotion of the cultural reputation of the hosts in the world community
- Introduction of new (and by implication, more modern) values and practices
- Revitalisation of traditional crafts, performing arts and rituals

*Main negative impacts*
- Debasement and the commercialisation of cultures
- Removal of meanings and values associated with traditional customs and practices when these are commodified for tourist consumption
- Increased tensions between imported and traditional lifestyles
- Erosion in the strength of local language
- New patterns of local consumption
- Risks of promotion of antisocial activities such as gambling and prostitution.

## Authenticity and commodification

Issues of authenticity and commodification reflect concerns over the manner in which indigenous cultures are used to promote and sustain international tourism. The success of modern tourist destinations will often depend upon the ease with which distinctive images of a place may be formed and marketed, and although images may be constructed around a variety of natural and non-natural elements, socio-cultural characteristics are especially important. There is, though, a tendency for processes of image-building by marketing agencies to misrepresent societies and cultures or to simplify them by characterisation. Herein lies the seed of a major problem, since the image obliges local people to present their traditional rituals and events, folk handicrafts, music and dance, religious ceremonies or sporting contests – all of which are capable of attracting tourists and forming a central element in their experience of the destination – in ways that accord with the image, rather than reality.

This is not to argue that tourist interests in culture are automatically detrimental as there is plenty of evidence to show how cultural places, artefacts and performances have been sustained through the interest and support of visitors. The experience of cultural tourism to Bali (see Box

7:2, pp. 168–9) illustrates the point. It is also true that in many tourist destinations, the tourist souvenir trade not only has contributed significantly to local economies but also has often helped to sustain traditional craftsmanship and helped to keep alive such traditions in communities that might otherwise have lost these skills and practices. This is especially true in emerging nations, but even in developed nations, tourist demand can be a vital element in sustaining local cultures. Many of the theatres, concert halls, galleries and museums of London's West End, for example, depend upon foreign tourists to maintain their economic viability.

However, as levels of demand increase and the composition of tourist markets moves towards mass forms of travel, there are very real risks of negative repercussions as cultural artefacts and performances become commodified (i.e. 'packaged' for convenient consumption or purchase by the tourist) and their authenticity is eroded. Such pseudo-events (as they are sometimes termed) generally share several characteristics:

- They are planned rather than spontaneous.
- They are designed to be performed or reproduced to order, at times or in locations that are convenient for the tourist.
- They hold an ambiguous relationship to real elements upon which they are based.
- Through time, they *become* authentic and therefore may replace the original event, practice or element that they purport to represent.

It is easy to see how tourist demand encourages these processes, and, on the positive side, it has been argued that by focusing tourist attentions upon staged representations of local culture – often within the comfort of the hotel lounge – the pseudo-event serves a valuable function in relieving pressures upon local communities and helps to protect their real cultural basis from the tourist gaze. But as it does so, it creates artificiality which detaches cultural elements from their true context. Tourists observing native ceremonies will usually lack the knowledge to comprehend the symbolism and true meaning of events, but there is a greater risk that through time the performers also lose sight of the original significance of a practice, and this alters its basis within the host culture. Likewise, the successful marketing of traditional objects as tourist souvenirs will alter their meanings or values, and where tourism markets develop, a tendency to increased dependence upon mass forms of production will marginalise the true craft worker. Mass production typically takes control over the development and sale of craft goods out

## Box 7:1

### Representation of native cultures in souvenirs: the case of Canada

For tourists to Canada, the purchase of souvenirs that depict native Indian and Inuit (Eskimo) peoples or their cultural forms is a popular and innocent practice. However, behind the practice lie very real problems of authenticity, appropriation of cultural images, misrepresentation of native cultures and the economic domination of the 'native' market by non-native producers.

The problems are particularly acute in the marketing of low-cost souvenirs, where the relatively small number of genuine native products are overwhelmed by mass-produced and often poorly manufactured imitations of native designs and artefacts or items where such designs and images have been applied to goods that form no part of traditional native cultures: tea-towels, key-rings, oven gloves, jigsaws and so forth. In many instances, these souvenirs are sold as representations of Canadian cultural identity and the specific origins in indigenous native culture are obscured.

A number of concerns over these practices have been voiced:

1 Native people rarely benefit economically from the use of their culture as production is dominated by non-native companies whose products undercut the prices of genuine native crafts.
2 Promotional practices often mislead tourists by loose and flexible use of terms such as 'authentic', 'original' and 'handmade'. In many cases, authenticity simply means that the item was manufactured somewhere in Canada or that it copies an original that was of native origin, rather than being the authentic product of a native Canadian.
3 Some souvenirs infringe copyright laws by reproducing the work of native artists in unauthorised forms. Beyond the economic loss to the artists, the quality of such work is often low and fails to represent adequately the skills of the artists.
4 Concerns have been expressed over the appropriation of native cultural identities to serve wider functions, particularly the promotion of a distinctive image of Canada to the international tourist. When native forms become symbols of the Canadian state as a whole, they tend to lose their more specific roles as symbols of a distinctive culture within that state.
5 Too many souvenirs misrepresent native cultures and lifestyles. This is a common problem within international tourism whereby native peoples are erroneously portrayed as leading an authentic, traditional and simple life that is no longer available to the modern world at large. The truth, however, is usually quite different.

The problems of cultural (mis)representation of native peoples in tourist souvenirs have formed one element in a much wider struggle on the part of native

Canadians aimed at enhancing their community identity and status within the nation as a whole. Limited attempts at legal regulation of the souvenir trade have made only slight impact and have fallen well short of the full protection of the collective rights of native peoples from the commodification of their culture by non-native commercial interests, for which some critics have called. The study notes, however, encouraging evidence of wider and more active promotion and marketing by native peoples of their own products. This not only directs economic gain from tourism to the native communities but also provides a means whereby artefacts are preserved within collective memories and sustains a sense of a distinctive native culture.

Source: Blundell (1993).

of local communities and into the hands of non-native producers. Box 7:1 summarises a study of some of the problems surrounding native craft production of tourist souvenirs in Canada that illustrates several of these points.

The extent to which authenticity issues and the commodification of culture by tourism is a real concern is very much a matter of opinion. The natural temptation is to decry the manner in which commercial tourism erodes and alters the cultural basis of host societies, but it is important to remember that culture is not static, it is dynamic and adaptive, and a vibrant society will constantly re-create and reconstruct its cultural basis. It is also a mistake to characterise native populations as passive objects of the tourist gaze since, as a number of authors have argued, many communities are actively constructing and promoting representations of their culture to attract the visitor, and many practices have thereby acquired new meanings and values. In this way, tourism should be conceived not as outside local cultures but rather as an integral part of ongoing processes of cultural formation.

## Moral drift and changing social values

A second area of general concern focuses upon the potential for contact between visitors and hosts to alter value systems and the moral basis to local societies, generally producing a drift towards adoption of more permissive or relaxed moral standards. To the local observer, the casual lifestyle of many tourists, their conspicuous consumption, their rejection (albeit temporarily) of normal strictures of dress and some elements of etiquette can create very diverse reactions amongst local people, although the strength of that reaction will depend upon the cultural distance between the host and the visitor (see Figure 7:1). Where differences are clear, the demonstration effect will tend to draw some elements in a society towards the alluring lifestyle that the tourists project (especially the young), whilst others (particularly older groups) will resist what are perhaps perceived as immoral forms of behaviour. (Such divisive tendencies have, for instance, been noted in several Mediterranean destinations, where the imposition of largely agnostic or atheistic North European tourists onto predominantly Catholic or Greek Orthodox communities with quite restrictive moral and social codes has been problematic.) In time, however, the natural processes of succession within the community will ensure that the value systems instilled in the young today are likely to become the norm for the society at large in the future. As a result, the effect of exposure to tourists may be to produce a drift towards changed moral and social values.

Once again, the extent to which this constitutes a 'problem' will depend upon the positions from which such changes are viewed, and whilst the temptation is to paint the tourist as the moral polluter of simpler, traditional societies, there are cases where roles are reversed and impacts are greater amongst the tourists than amongst the hosts. Visitors to Scandinavian countries or the Netherlands, for example, may find prevailing moral codes that adopt rather more tolerant attitudes to social issues such as sexuality and bisexuality, drug use or prostitution than those that prevail at home. Under these circumstances it is the tourists, not the hosts, who are likely to experience a challenge to their traditional moral codes and behaviours.

An outline review of the literature will reveal a core of common social issues that are routinely linked with tourism. These include tendencies for tourism to be associated with increased incidence of gambling, prostitution and certain types of crime, together with a rather more subtle impact upon religions within host countries.

Gambling, prostitution and crime are frequently interlinked, both through organisational structures in which ownership of casinos and that of brothels are often vested in the same hands and sometimes financed by profits from criminal activity, and through spatial proximities in which clubs, casinos, bars and brothels cluster to produce 'red light' or 'entertainment' districts. London's Soho and Amsterdam's Warmoesstraat are examples. The links to tourism are, however, much less clear.

The hedonistic character of many forms of tourism will foster some interest in activities such as gambling or prostitution, although interest will occur selectively and, with the occasional exception of resorts such as Las Vegas, only small minorities of visitors will actually indulge in these activities. The sensitivities that surround these practices (particularly prostitution) have resulted in there having been very few empirical studies of the role of tourism in their development, but those studies that have been completed tend to confirm the view that tourism promotes existing practices, rather than causing activity in a direct sense. Thailand has developed a dubious reputation for sex tourism, yet it is clear that prostitution was an established element in local urban subcultures in Thailand long before the arrival of tourists. The primary effect of tourism seems to have been to encourage the addition of a tier of expensive, elite young women to meet the new demands of the tourist market, although there is also some evidence to link tourism with newer forms of sexual exploitation, especially involving children.

Similarly, links between tourism and local crime are not always clear and consistent. Visible differences in levels of affluence between visitor and host are sometimes held to account for increases in the incidence of robbery and muggings, especially as tourists moving around strange locations, and often unable to distinguish 'safe' from 'unsafe' areas, are easy targets for streetwise criminals. Studies of destinations in places as varied as Queensland, North Carolina, Hawaii and New Zealand show links between tourism development and increased rates of burglary, vandalism, drunk and disorderly behaviour, sexual and drug-related offences and soliciting by prostitutes (which is a criminal activity in many countries), but statistical linkages do not necessarily mean that tourism *causes* such activity. The normal practices of tourists create conditions and environments in which many forms of crime will flourish, but except in situations where the tourists themselves are perpetrators of crime (as, for example, in the rising incidence of drunken and violent behaviour by young British tourists in Spanish Mediterranean resorts or, less commonly, the smuggling of drugs or other illegal goods by visitors), tourism cannot introduce crime to a host society. The tendency must already exist, albeit, perhaps, in a latent form.

The moral value systems in many societies are rooted (if only distantly) in religious beliefs and practices, so the capacity of local communities to resist changes to moral codes may be partly dependent upon the strength of the religious basis to daily life. The links between tourism and religion have changed through time in some interesting ways. Religion was, and still is, a basis for particular forms of tourism, but whilst in many societies – especially in the developed world – belief in religion has been eroded in the face of growing agnosticism and atheism, religious sites have become an increasingly popular object of the tourist gaze, even if people do not subscribe to the beliefs that such places represent. There is no doubt that worshippers at the great Anglican cathedrals of England are greatly outnumbered by the millions of tourists who come simply to view the buildings.

This is a potential source of conflict when the practices of the devout are directly compromised by the idle curiosity of the masses. For most tourists, religion has become entertainment, typically in the form of casual inspection of religious sites or the observation of religious ceremony. For the worshipper, or the participant in a religious ceremony, the place or the event has quite different meanings and may be a source of profound spiritual, moral and psychological support. Any devaluation of the experience, therefore, whether it be through the commodified performance of religious spectacles for tourist consumption or irreverent behaviour on the part of tourists towards religious places or practices, may be deeply disturbing. Yet as before, the effect of such encounters will be unpredictable. On the one hand, it may serve to reinforce local adherence to religiously based practices and values, strengthening a sense of local cultural identity. Equally, it may erode the position of religion within societies, altering the meaning and symbolism of ceremonial and events and opening the way towards wider processes of social and cultural change.

## New social structures and empowerment

The composite effect of many of the socio-cultural changes that have been associated with tourism may eventually lead to significant shifts in local social structures and new patterns of social empowerment. As before, effects are likely to be most pronounced where tourism brings together hosts and visitors from contrasting socio-economic traditions, but where such differences are marked, repercussions could be significant.

Changes result through a number of pathways, but two are worth emphasis. First, tourism creates new patterns of employment and opportunities for work amongst groups who, in traditional societies, may not normally work for remuneration, for example women. It has been argued that in many agrarian societies, the arrival of tourism has had particularly beneficial effects for young people who gain employment in the industry, enabling new levels of financial independence, partial or total release from the traditional social controls of their elders that normally exist within extended families, and new choices in matters such as place of residence or selection of marriage partners. As we have seen in Chapter 4, tourism creates particular opportunities for employment for women, and it has been argued that one of the beneficial effects of tourism has been to help the liberation of women from traditional social structures, to provide the independence that comes with a personal income, and to promote, through time, more egalitarian social forms and practices.

Such social empowerment may also arise through the second key process: language change. Language is a significant defining feature of a society. It provides identity but, more significantly, it underpins social patterns by simply defining who talks to whom. However, because international tourism is generally conducted through one of a very few languages that have world-wide usage, most typically English, the normal expectation of the tourists is that the hosts will have at least some grasp of their language. Expectation is often reinforced by practice. Foreign ownership of tourism developments may impose a new language as the norm for business purposes, whilst training in the hospitality industries will also strive to give personnel some grasp of languages that they are likely to encounter. But as with employment, so the acquisition of new language skills empowers people in several significant ways. It provides wider access to globalised media and the influences that the media convey; it makes easier the possibility of migration in search of employment or improved prospects; and it alters the status of the individual within their home society through the acquisition of a powerful skill that others may lack.

The social empowerment that comes with employment or the adoption of new languages is best envisaged as operating at the individual or group level. But occasionally, whole communities and cultures become empowered through the development of tourism and its integration into local socio-cultural development. Box 7:2 outlines an example of this process on the Indonesian island of Bali.

## Box 7:2

### *Cultural tourism and empowerment: the case of Bali, Indonesia*

Tourism to Bali was initially a product of the Dutch colonial era, when the cultural attractions of the island were first realised. The unusually expressive culture, rooted firmly in Buddhism and manifest in a remarkable range of religious sites, ceremonies and visually striking performing and visual arts, was seen as a particular attraction to overseas visitors. When Bali became part of Indonesia in 1958, the promotion of Balinese tourism – primarily as a means of earning foreign revenue – was reaffirmed. The first tourism master plan (1971) proposed a series of enclave developments which, by 1988, had been extended to cover fifteen locations, mostly concentrated on the south coast.

The reaction of the Balinese to the active development of tourism was initially mixed. The economic gains were largely welcomed but there were concerns over the extent to which the cultural heritage of the island would be protected. The Indonesian government was felt by some to be appropriating Balinese culture as a key component in attempts to forge an Indonesian 'national' culture, and much of the rationale for the initiative was seen as political – an attempt to project a positive image of Indonesia to the international community, in order to distract attention from internal tensions and extensive abuses of human rights.

However, the Balinese authorities soon realised that the new strategic role of tourism actually provided a political and economic lever that they could use to advantage in their dealings with the Indonesian government. The Balinese adopted a conscious policy of:

- fostering cultural tourism as a major attraction;
- developing new resorts to spread economic benefits more evenly;
- using the international recognition of Bali to strengthen the political position of Bali within the Indonesian state.

By turning Balinese culture into a major resource, the economic benefits would follow and, at the same time, strengthen the Balinese case for preserving and protecting Bali's distinctive cultural heritage – *empowering* the Balinese in a political sense and helping to create new levels of self-assurance in their dealings with the national government.

Bali therefore presents an unusual situation in which, rather than tourism and cultural preservation being viewed as opposing interests, the development of tourism and the preservation of culture are seen as self-reinforcing. The experience of cultural tourism has actually encouraged the Balinese in the belief that rather than being an agent of cultural pollution and change, tourism has provided the political and social empowerment that has led to a social and cultural renaissance. The character of that renaissance has perhaps been selective, since the Indonesian authorities have consciously promoted only those aspects of

Balinese culture that they believe reflect most positively on the nation at large. However, the process has still had a significant impact in reviving artistic traditions and reinforcing the view of Bali as a distinctive society, with a clear sense of cultural identity, existing within the Indonesian state.

Source: Picard (1993, 1995).

## Conclusion

The impacts that tourism brings to host societies and cultures are remarkably diverse and often inconsistent in their effect, reflecting the many different ways under which people travel and the variations in local conditions that they encounter. In some situations, where cultural distances between hosts and visitors are slight, socio-cultural effects of tourism are minimal. Elsewhere, changes are more significant, and although the tendency in many of the discussions of socio-cultural relationships between hosts and visitors is to emphasise the negative, the preceding sections have attempted to show that there are often significant and tangible benefits from encounters between tourists and local people too. In particular, the example of Bali shows how by integrating tourism into wider programmes for socio-cultural development, tourism can actually become an agent for empowerment and help to assert distinctive local identities in a world that is increasingly shaped by global processes.

It is also important to re-emphasise the point that societies and cultures are not fixed entities, nor are hosts the passive receivers of the stimuli to change that the visitor may bring. Society and cultures evolve constantly, in response to a wide range of external and internal influences – one of which is clearly international tourism. But it must be remembered that tourism is one of many such influences, and disentangling the effects of tourism from those of, *inter alia*, multinational corporations, international political organisations, global media, aid and charitable groups, and cultural exchange and educational programmes is probably an impossible task.

## Summary

The impacts of tourism upon societies and cultures are often reflected in imprecise ways, but few doubt that through processes of acculturation or the demonstration effect, tourism has the power to alter socio-cultural structures in destination areas, even though the precise forms of such effects are often uncertain and spatially variable. A diversity of factors may be held to account for such variation, including the nature and scale of host–visitor encounters, the cultural 'distance' between the different groups, and the stages of tourism development that have been attained. There is also a range of possible effects, including issues of cultural commodification and misrepresentation, the introduction of new moral codes, or the promotion of new social value systems. However, whilst the tendency is to represent tourism as a form of socio-cultural 'pollution', there is a growing body of evidence to show that processes of cultural influence are often two-way, and further, that positive socio-cultural impacts may be initiated through host–visitor contacts.

## Discussion questions

1  How convincing is the concept of the demonstration effect as an explanation of tourist impacts upon host societies?
2  Why has globalisation of tourism sometimes produced a reassertion of local values and practices?
3  Can tourism, in isolation, actually destroy host societies and their cultures?
4  Why does the commodification of tourism threaten the authenticity of cultural representations?
5  In what ways can tourism promote cultural empowerment?

## Further reading

Excellent general discussions of the impact of tourism upon societies and cultures may be found in:
Mathieson, A. and Wall, G. (1997) *Tourism: Change, Impacts and Opportunities*, Harlow: Longman
Murphy, P.E. (1985) *Tourism: A Community Approach*, London: Routledge: 117–151.

and a particularly well-balanced critique in:
Ryan, C. (1991) *Recreational Tourism: A Social Science Perspective*, London: Routledge: 130–166.

A broader and highly perceptive discussion of the general socio-cultural context of tourism is provided by:
Krippendorf, J. (1987) *The Holiday Makers*, Oxford: Butterworth Heinemann.

whilst an interesting collection of case studies written from a broadly social anthropological perspective is contained in:

Lanfant, M.-F., Allcock, J.B. and Bruner, E.M. (eds) (1995) *International Tourism: Identity and Change*, London: Sage.

For a comprehensive examination of cultural tourism in Europe see:

Richards, G. (ed.) (1996) *Cultural Tourism in Europe*, Wallingford: CAB International.

In addition to case studies cited in the text, useful specific examples of tourism impacts on society and culture may be found in:

Getz, D. (1994) 'Residents' attitudes towards tourism: a longitudinal study in Spey Valley, Scotland', *Tourism Management*, Vol. 15 No. 4: 247–258.

King, V.T. (1993) 'Tourism and culture in Malaya'. In Hitchcock, M., King, V.T. and Parnwell, M.J.G. (eds) (1993) *Tourism in South East Asia*, London: Routledge: 99–116.

Mercer, D. (1994) 'Native peoples and tourism: conflict and compromise'. In Theobald, W. (ed.) (1994) *Global Tourism: The Next Decade*, Oxford: Butterworth Heinemann: 124–145.

Teo, P. (1994) 'Assessing socio-cultural impacts: the case of Singapore', *Tourism Management*, Vol. 15 No. 2: 126–136.

# 8 ▸ Inventing places: cultural constructions and alternative tourism geographies

Places, and images of places, are fundamental to the practice of tourism. The demand for tourism commonly emanates from individual or collective perceptions of tourist experiences that are usually firmly rooted in associations with particular places, whilst the promotion and marketing of tourism depends heavily upon the formation and dissemination of positive and attractive images of destinations as places. Tourism therefore maps the globe in a distinctive, though highly subjective, manner, and one of the ways in which we may view the geography of tourism is as a visible manifestation of perceptions and images of what constitute tourism places. However, as those perceptions and images are recast and re-formed – in response to changing public expectations, tastes, fashions, levels of awareness, mobility and affluence – new tourism geographies emerge, overlying or even superseding previous patterns as different forms of tourism promote new areas of interest. This chapter explores some of the ways through which such tourism geographies are formed and in concluding the volume as a whole, points up some of the new directions in which tourism appears to be moving.

## Inventing tourism places

The construction and subsequent consumption of tourist places is essentially a socio-cultural process. The initial identification of places to which tourism may be drawn reflects an appraisal of resources that is located in cultural evaluations, and the physical development of tourism places typically depends as much upon social and institutional structures and organisations as it does upon the more tangible impacts of (say) innovations in transport technology. Hence, the original growth of sea bathing resorts in eighteenth-century England mirrored key shifts in health practices and beliefs, whilst the later development of mountain tourism in Alpine Europe owed its impetus to the alternative views of landscape that grew out of the new taste for the picturesque and the romantic that was popularised in the first decades of the nineteenth century. Later still, the growth of mass forms of Mediterranean tourism only became really established with the fashion for sunbathing from about 1920. Railways (and later aeroplanes) may have provided a physical mechanism for moving tourists in large number to new destinations, but it required the transformations in the social organisation of tourism (for example, the guided tour and later the packaged holiday) and the importation of holidaymaking into popular culture to realise that potential to the full.

## The tourist gaze

The spatial pattern of tourism depends very strongly, therefore, upon how we view places – an idea that has been most usefully conceptualised by the sociologist Urry in his notion of the 'tourist gaze'. The concept of the gaze is valuable because it helps us understand the processes both of construction of tourist places and of their consumption, whilst the metaphor of visualisation that is implicit in the term 'gaze' is central to comprehending modern tourism practices and their associated meanings. Tourism is a strongly *visual* practice. We spend time in advance of a tourism trip attempting to visualise the experience by examining guide books and brochures, or in anticipatory day-dreams; we often spend significant parts of the trip itself engaged in the act of sightseeing in which we gaze upon places, people and their artefacts; and we relive experiences as memories and recollections, aided by photographs or home video footage that we have consciously taken to act as visible reminders of the trip. However, the entire process of visualisation, experience and recall is socially constructed and strongly mediated by 'cultural filters'. We gaze and record places in a highly selective fashion, disregarding some places altogether and, from the remainder, removing the unappealing or the uninteresting. In the process, we are inventing (or reinventing) places to suit our purposes. The gaze is also a detached and superficial process, as the term itself suggests. This superficiality increases the role of cultural signs within the invention and consumption of tourist places – not signs in the literal sense of directional indicators, but figurative signs: places or actions that represent, through simplification, much more complex ideas and practices. So for the tourist, a prospect of a rose-decked thatched cottage may come to represent or embody a much wider image of 'olde England' and the lifestyles and practices that mythologies associate with the rural past. For some writers, tourism has become an exercise in the collection of such signs – the postcards and the holiday photographs from the great tourism sites of the world conferring a status upon the individual, the true mark of the modern (or post-modern) tourist.

The construction of the tourist gaze is also inextricably bound up in the notion of contrast or difference. Tourism, or at least those leisurely sectors of tourism, is widely conceived as an opposite to work, and the practice of travel takes the tourist away from the familiarity of places of employment or residence and into places that have been consciously selected as providing varying levels of contrast to the familiar. This helps to explain the clear tendency for tourism geographies to change through time, through the quest to discover (or invent) new tourism places to replace established locations that have become unacceptably familiar. This may result in the tourist gaze being focused upon new destinations or perhaps upon elements in existing destinations that had not previously been a part of the tourist circuit and which therefore possess the valued cachet of originality. So, Brighton and Torbay are replaced by Biarritz and St Tropez, whilst the seasoned tourist to Paris, no longer simply content with views of the Eiffel Tower, may now sign up for guided visits to the city's nineteenth-century sewers.

The manner in which we gaze upon tourist sites (and sights) is partly a product of our own social, educational and cultural backgrounds and partly a result of the systematic production and presentation of tourism places within the media in general and the travel industry in particular – a form of 'professional gaze'. Film, television, magazines, travel books and advertisements constantly produce and reproduce objects for the tourist gaze. This is an enormously powerful influence that infiltrates the subconscious of everyday life, creating new patterns of awareness, fuelling desires to see the places portrayed and instilling within the travelling public new ways of seeing tourism destinations. Research suggests that most visitors' perceptions of tourism places are often vague and ill-formed, unless those perceptions have been sharpened through previous experiences. Hence there is clear potential for marketing and promotional strategies to shape both the character and the direction of the tourist gaze.

## Promoting tourist places

Surprisingly, perhaps, relatively little work has been conducted upon the role of advertising in the cultural representation of tourist places, despite the lengthy – if not always honourable – history of the practice. Some of the first promoters of tourism places were the railway companies, which, in their efforts to secure a commercial market, produced some enduring images of places. Visitors to contemporary Torquay, for example, are still welcomed to the 'English Riviera' – a conception (originally in the more spatially specific guise of the 'Cornish Riviera') of the Great Western Railway in the early 1900s. Under the distinctly patriotic slogan 'See your own country first', it exhorted potential travellers to explore the delights of distant and exotic Cornwall in preference to Italy, with which it drew direct parallels in terms of the mildness of climate, the natural attractiveness of its (female) peasantry and even the shape of the two lands on the map, albeit with some cartographic licence in the case of Cornwall. In so doing, the railway promoters fed off such limited perceptions of Cornwall as may have existed, but primarily they invented an image that was then reinforced through associated guidebooks and literature that presented Cornwall as some form of distant, yet still accessible, Arcadia.

This tradition of creative promotion of tourist places has continued to the present, and content analysis of contemporary tourist brochures reveals texts that are often unashamedly escapist in their tone and which, when combined with photographic representations, emphasise difference; excitement; timelessness or the unspoilt; tradition or romance – according to the perceived market at which the publicity is aimed.

Such creative constructions of tourist places are most prevalent, of course, in the representation of foreign destinations, where fewer people will have had the direct experience needed to balance the claims of the brochures and the guidebooks. Messages are often subtly encoded. A recent study of a cross-section of British travel brochures promoting foreign places found, for example, that 25 per cent of illustrations showed only empty landscapes and, especially, beachscapes (reinforcing

ideas of escape); that pictures showing tourists were nine times more
common than pictures of local people (reinforcing notions of exclusivity
and segregation); and that written text placed overwhelming emphasis
upon qualities of naturalness (as an antithesis to the presumed artificiality
of the tourists' routine lives) and the opportunities for self-(re)discovery.
Only occasionally were senses of the exotic conveyed by use of images
of local people, whilst reassurance that the experience of difference
would not be so great as to be disorientating and unpleasant was
provided by pictures showing familiar (though culturally displaced)

**Figure 8:1** *Imagined tourism 'countries' in England*

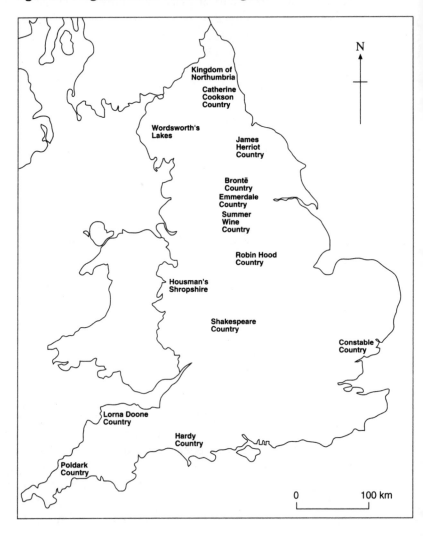

items – typically as background elements. Examples of the latter might include 'English-style' pubs in the Spanish package tour resorts or, most ubiquitously of all, glimpses of the familiar red emblem of the Coca-Cola Company.

Promotional material that presents selective representations of realities is, of course, to be expected. What is more interesting, perhaps, is the emerging trend in some sectors of tourism towards promotion of places that do not actually exist or which are entirely imagined reconstructions of a locality. This has been nicely exemplified in England and Wales by the popular practice within regional and local tourism boards of appropriating legendary, literary or popular television characters or events to provide a form of spatial identity to which tourists will then be drawn (Figure 8:1). Some are well established. The term 'Shakespeare Country' to designate the area around Stratford-upon-Avon dates back to railway advertising of the 1930s and, along with similar descriptions of the Lake District as 'Wordsworth Country' or Haworth as 'Brontë Country', possesses some grounding in the real lives of individuals. 'Robin Hood Country' is more problematic given the uncertainties surrounding the actual existence of Robin Hood. However, descriptions of parts of Tyneside as 'Catherine Cookson Country' or the Yorkshire Dales as 'Emmerdale Farm Country' or Exmoor as 'Lorna Doone Country' take the process one stage further removed. They confuse reality with fictional literary and television characters or locales, and tourists are thereby confronted by a representation of what is already a representation. It is then only a short step to the totally artificial worlds of Disney in which cartoon characters – albeit in the form of employees in costume – step into the sunlight of Anaheim or Orlando to be photographed with the tourist.

## Commodification and pseudo-realities

Such practices represent commodification of tourism in one of its most overt forms. This is the tourism industry constructing a product and marketing it as an inclusive and convenient experience of another place. It draws selectively upon the real nature of places and presents only those elements that will appeal to the market segments at which the holidays are directed. But given the alacrity with which tourists consume such commodified and invented places, a question is raised over the significance, or otherwise, of 'real' experiences of place.

The problem for providers of tourism is that having constructed specific images of peoples and places in order to draw the visitor, it is obligatory for destinations to match the images that are projected. The tourist must confirm his or her expectations, otherwise the return visits, or the visits of others made on the basis of personal recommendation (one of the most important methods for disseminating knowledge of tourism places), will not occur. In this way, tourist images tend to become self-perpetuating and self-reinforcing with the attendant risk that, through time, tourist experiences become increasingly artificial.

The argument that tourist experience is founded on artificial rather than real situations is an idea that has been debated since the early 1960s, when a number of scholars (most notably Boorstin) attempted to argue that the traveller does not experience reality but thrives instead upon 'pseudo-events' – commodified, managed and contrived forms of provision that present a flavour of foreign places in a selective and controlled manner. This is evidenced in several distinct directions.

1 *The physical isolation of the tourist from the host environment.* This tendency receives its most obvious expression in the enclavic forms of resort development in the Third World (see Chapter 4), where visitors are provided with the familiar creature comforts that may literally have been imported from their place of origin, set within a physical environment that has been deliberately contrived to reflect popular images of what an exotic location should be like. But many forms of tourism place visitors in what has been termed an 'environmental bubble': a protective cocoon of Western-style hotels, international cuisine, satellite television, guidebooks and helpful, multilingual couriers – 'surrogate parents' that cushion and, as necessary, protect the tourist from harsher realities and unnecessary contacts. As a result, the tourist gaze is often akin to gazing into a mirror. We construct tourism places to reflect ourselves, rather than the places we are visiting.

2 *Cultural imposition.* The powerful expectations of tourists often impose particular forms of development and provision upon host communities. Many of us, quite illogically, expect a home-from-home experience, even in foreign lands, and the necessity for local providers to match those expectations inevitably changes the nature of the places that we visit. In the most extreme forms of this phenomenon, places actually begin to lose their sense of identity – they become *placeless* and quite indistinct from other tourist places, and quite unrepresentative of the realities of indigenous places. The mass tourist

resorts of the Spanish coast, for example, commonly present a bland, placeless uniformity that says little about the 'real' Spain that exists often only a few miles inland.

3 *Staged events.* One of the many ironies in international travel is that a primary motive is exposure to foreign culture and custom, yet this is often met through a quite artificial purveyance of supposed custom, whether via the sale of inauthentic souvenirs (see Box 7:1) or via staged events or places. Sanitised, simplified and staged representations of places, histories, cultures and societies match the superficiality of the tourist gaze and meet tourist demands for entertaining and digestible experiences, yet they provide only partial representations of realities.

## New tourism directions

In some ways, commodification and the pseudo-realities represent an inevitable product of modern tourism if mass markets are to be effectively served, but the visible shift from a natural to an artificial basis to tourism and the lack of authenticity that surrounds many contemporary tourism practices have raised concerns. For some writers (especially MacCannell), tourists embody a quest for authenticity – the act of travelling and the shifting focus of the tourist gaze reflecting a search for an authenticity that many tourists from urban, industrialised countries seem no longer able to detect in their own routine lifestyles. For others, issues of authenticity are much less important than the quest for difference.

The quest for difference has become one of several strands within an emerging pattern of 'new' tourism – alternative tourist geographies that have been variously designated as 'post-industrial', 'post-modern' or 'post-Fordist'. The new tourism industry is characterised by flexibility, by segmentation and by the development of new forms of customised experience that bring a myriad of new choices to tourists and thus offer stark contrasts to the mass, standardised and packaged forms of 'old' tourism. Table 8:1 summarises one perspective on changing patterns of consumption and the ways in which tourist patterns reflect such shifts, here conceptualised in terms of a Fordist/post-Fordist dichotomy. (Fordism, which is derived from the production-line philosophy of the motor manufacturer Henry Ford, is a description given to the hypermodern patterns of mass, standardised forms of production and

**Table 8:1** *Tourism and post-Fordist forms of consumption*

| Post-Fordist consumption | Tourist examples |
|---|---|
| Increased significance of consumers and consumer choices | Rejection of some mass forms of tourism (holiday camps; cheap packages) and increased diversity of preferences |
| Greater volatility of consumer preferences | Fewer repeat visits and proliferation of alternative attractions |
| Increased market segmentation | Multiplication of holiday types |
| More new products with shorter product life | Rapid turnover and change in popularity of sites in response to fashions |
| Increased preference for non-mass forms of production and consumption | Growth of alternative tourism |

Source: Adapted from Urry (1995).

consumption that prevailed between approximately 1920 and 1970. Post-Fordism, therefore, emphasises contrasting patterns centred on flexible production and enhanced consumer choice.)

The advent of new forms of tourism does not, of course, mean that old tourism in its mass, packaged forms is disappearing. There remain many millions of people committed to such travel who derive sufficient rewards and pleasures from the experience to sustain mass tourism for the foreseeable future. What is changing is the development of new market segments comprising groups seeking the out of the ordinary, groups that we may expect to see forming alternative tourism geographies that reflect this quest for difference. These trends are revealed within the increasing popularity of, *inter alia*:

- cultural tourism, which has become a major factor shaping tourism patterns in Europe, refocusing attentions upon major cities both as established tourist destinations – for example, Paris, Amsterdam, Florence and Vienna – and as new centres of culture – for example, Glasgow;
- heritage tourism, which is especially important in countries such the UK where a wealth of heritage-related sites survive, often in places that represent new locations for tourism;
- theme parks, which are arguably the outstanding example of the artificial construction of post-modern tourist spaces and are fast

becoming major centres of tourist consumption in countries such as
the USA, Britain and Japan;

- adventure tourism, which frequently takes tourists to exotic and
  unexplored places where notions of difference are often especially
  prominent; for example, trekking in the Himalayas;
- eco-tourism, which attempts to combine principles of sustainability in
  developing new forms of responsible travel to natural areas,
  particularly to Third World destinations such as Costa Rica, Ecuador
  and Nepal.

The new tourists – 'post-tourists' in the minds of some writers – embody
a new spirit of playfulness in tourism as a dominant mode of experience.
They are not deceived by the pseudo-realities of contemporary tourism
but are happy to accept such constructions at face value as an expected
and valued part of new forms of experience.

The broad effect of these demands has been twofold: the reinvention of
existing tourist places and, much more widely, the establishment of new
ones. We will explore the first theme by briefly revisiting a traditional
seaside resort and then examine the latter theme more fully by reference
to two contrasting examples: the growth of heritage tourism in Britain
and the growth of theme parks, particularly in the USA and Japan.

## Reinvention of existing tourist places

The reinvention of places is a perennial theme within tourism
development and many different types of destination have periodically
been required to adjust the nature of what they offer to the visitor in order
to keep abreast of evolving tastes and fashions. This is most visibly
evidenced in the changing role of traditional seaside resorts. In Chapter 2
(Box 2:1), we visited Brighton in its formative years and saw how the
resort was 'invented' through a particular combination of scientific
advocacy, opportunism, patronage, fashion, technology and new social
organisation. Revisiting Brighton today, we find a resort that is busy
reinventing itself, partly in an attempt to retain a tourism sector in the
face of growing competition from foreign places, and partly to build new
images of the town in order to attract alternative populations and
enterprises – particularly in commercial services and higher education.
Tourism policies are especially interesting since they reveal a conscious
attempt to define a new image for Brighton that actually mirrors its
distant past, replacing the vulgar, even seedy, version of Brighton that

had developed in the 1950s and 1960s with a more stylish and up-market model that consciously refers to the period of Regency Brighton. Some of the reinvention of Brighton has centred on conference tourism, which has become a major user of the elegant seafront hotels with more than 1,200 conferences hosted in the town in 1994. But more significant has been the growth of heritage tourism, a process in which the townscapes that tourism created in the nineteenth century have now become the object of a new tourist gaze. Urban conservation areas have mushroomed; the Old Town has been pedestrianised and, in the process, the streetscapes have been reconstructed by the implantation of street furniture (signs, seats, litter bins, etc.) designed to look old; whilst the Brighton Festival, once a low-key, local event, has been developed to form a major attraction aimed at high-spending cultural tourists. The boisterous seaside entertainments that characterised the Brighton of the recent past are now closely contained by local planning controls, and whilst the disused West Pier corrodes away and falls into the sea – a potent symbol of the passing of one tourism era – millions of pounds are being reinvested in restoration of King George IV's Royal Pavilion as a focal point for the new heritage tourist.

## The establishment of new tourist places

### Heritage tourism

Heritage tourism (in which the term 'heritage' is taken to refer to places, objects or ideas that are deemed to be of value or importance and which have been passed from one generation to the next) has, in recent years, developed as one of the main sectors in the establishment of new spatial patterns of tourism, particularly in countries such as Britain. Brighton represents an example of an existing tourism place using its past to sustain its future, but heritage tourism has also been more widely responsible for the introduction of tourism into localities that previously held no pretensions to be tourist places. In Britain (which has a stronger heritage tourism tradition than most), the castles, great houses, museums, galleries and historic towns that have provided the basis of one set of tourist geographies for some time have been supplemented by new tourist spaces that centre upon industrial cities, ports and the working countryside – places which have adapted (and adopted) their pasts as a means to attract visitors.

The dramatic expansion of heritage tourism invites both description and explanation. The practice of gazing on the past is not new. The Grand Tourists of the seventeenth and eighteenth centuries focused much of their attention on sites and artefacts that we would today classify as 'heritage', whilst in the nineteenth century, popular rail excursions ferried industrial workers as day trippers to places such as Windermere to enjoy the landscape heritage of the Lake District. What has changed is the scale, diversity and extent of heritage-based attractions in tourism, and this is evidenced in Table 8:2, which summarises a typology of heritage attractions. Since the mid-1970s, Britain has acquired over 1,000 new registered museums, an additional 210,000 listed buildings (i.e. buildings designated as being of architectural importance), 5,000 new

**Table 8:2  *A typology of heritage tourism attractions***

- *Natural history and science attractions*: including nature reserves and trails; zoos, aquariums, wildlife parks and rare breeds; technology centres; scientific museums; geomorphological or geological sites (caves, gorges, cliffs or waterfalls)

- *Agricultural and industrial attractions*: including working farms and farming museums; quarries and mines; factories; breweries and distilleries; museums of industry

- *Transport attractions*: including transport museums; working steam railways; canals and docks; preserved ships; aircraft and aviation displays

- *Socio-cultural attractions*: including historic sites; museums of rural or industrial life; museums of costume

- *Built attractions*: including stately homes; religious buildings (cathedrals, churches, shrines)

- *Military attractions*: including castles, battlefields, naval dockyards and military museums

- *Landscape attractions*: including historic town- and villagescapes; national parks; heritage coastlines and seascapes

- *Artistic attractions*: including galleries; theatres or concert halls and their performances; art festivals

- *Attractions associated with historic figures*: including homes or working places of writers, artists, composers, politicians, military leaders or leaders of popular culture

Source: Adapted from Prentice (1994).

conservation areas and a further 5,400 scheduled ancient monuments. Over 2,000 historic buildings and monuments are regularly open to the public, and on a recent Heritage Open Day (when sites not normally open received visitors), some 670,000 people took advantage of the opportunity, part of an overall total of an estimated 58.3 million heritage visits made in the same year – 1996. Recent trends (Figure 8:2) suggest that although prone to year-on-year fluctuations, the market trend for this form of tourism is still upward.

Explanation for the popularisation of the past in contemporary tourism needs to take account of several factors. First, it is argued that interest in the past is something inbuilt in human nature, and that for many people the remains of the past provide a sense of security and continuity that will be amplified whenever the present becomes uncertain. The concept of nostalgia is important here. Nostalgia was initially recognised as a physical condition (first seen in sailors on long voyages of discovery); we now call it 'homesickness'. The uncertainties surrounding modern lives, so it is argued, produce an equivalent of the sailor's homesickness, but here manifest in sentimental attachments to a past that is variously perceived as safer, more secure and more predictable.

This links to a second idea. One of the abiding themes of contemporary life – and strongly evident in tourism – is globalisation. Yet

**Figure 8:2   *Market trends in British heritage tourism, 1984–96***

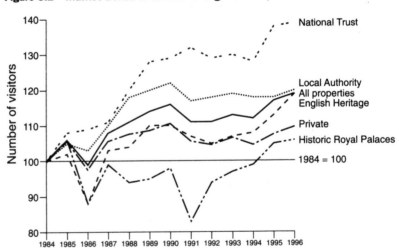

Source: British Tourist Authority (1997) *English Heritage Monitor 1997.*

paradoxically, as our lives become more influenced by globalised processes, so resistance to globalisation becomes more firmly embedded in the reassertion of local places, histories and cultures. Out of this reassertion of the local has arisen a popular preservation movement that has actively reclaimed and restored a wide range of places and associated artefacts as cultural mementoes and, through their active promotion, developed their use as objects of the tourist gaze.

This process has been aided by a third factor. The rapid deindustrialisation of Britain in the 1970s and 1980s – as happened in a number of industrialised economies – created a valuable legacy of redundant sites and buildings that provided natural homes for the new heritage industry. As working industry disappeared, people became increasingly interested by its memory and by the forms of life and practices associated with it. This trend, in particular, has been responsible for much of the diversification within heritage tourism, focusing popular attention on a wide range of industrial museums, restored mills and factories, working steam railways, canals and dockland, coal mines and slate quarries – places that represent new sites of consumption on the tourist map. The commodified, selective and sanitised ways in which such places tend to present their views of the past have drawn predictable criticism over a lack of authenticity and a tendency to glamorise, yet the majority still convey a valuable sense of the past that many visitors would otherwise struggle to achieve through other media and are consciously valued for this quality.

There have also been important social shifts that have helped to promote heritage tourism. The collective weakening of traditional working classes in many Western economies, which has been partly responsible for the decline of older seaside towns, has been countered by the emergence of new professional and quasi-professional service classes whose lifestyles and aspirations are much more attuned to the enjoyment of cultural and heritage-based attractions. However, one of the great strengths of heritage tourism is its breadth. The past means different things to different people, and whilst there is still a strong market for 'high' cultural forms of the kind expressed in traditional museums, art galleries, concert halls and theatres, the post-industrial era has been associated with wider promotion of 'low' or 'popular' cultural forms that are far more accessible to people who lack the cultural capital to enjoy – or even make sense of – the high forms of heritage. So the new museums of industry, of transport, or of rural life, for example, succeed because they offer memories to which ordinary people can relate.

## Box 8:1

### *Development of heritage tourism in Bradford, England*

Bradford Metropolitan District, with its population of 480,000 people, is located on the western fringes of the West Yorkshire conurbation (Figure 8:3). Bradford is a product of the Industrial Revolution of the nineteenth century, and its economic base until very recently was founded on textiles and engineering,

**Figure 8:3  *Bradford: location and major tourism attractions***

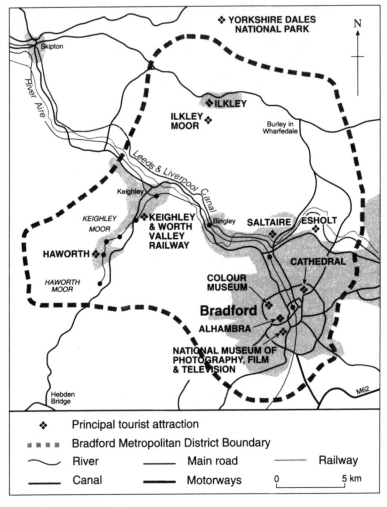

Source: Adapted from Davidson and Maitland (1997).

factors which had contributed to a popular image of the place as a grim and grimy northern city, plagued by poor housing and amenities and suffering extreme levels of unemployment. By the 1980s, inevitable decline in traditional industries posed major challenges to attempts to restructure and revive the area.

Surprisingly, the new strategies that emerged to address these problems included proposals for the promotion of tourism, using several established or potential attractions as a basis. These included:

- an existing stock of hotel bed spaces associated with the city's commercial activity;
- a superb Victorian industrial heritage, including the model community of Saltaire;
- proximity to the Yorkshire Dales National Park and other scenic areas such as Ilkley Moor;
- the village of Haworth (which lies in the Metropolitan District) and which is the centre of 'Brontë Country';
- the Keighley and Worth Valley steam railway;
- locational settings for the popular TV soap *Emmerdale*.

To these existing attractions, the city was able to add further to its tourism stock, including the National Museum of Photography, Film and Television (which was established in 1983 as a joint partnership between the Science Museum and the city council); the restored Alhambra Theatre (one of Britain's best-preserved

**Table 8:3** *Major tourism attractions in the Bradford Metropolitan District, 1994*

| Attraction | No. of visitors |
| --- | --- |
| Haworth village | 1,000,000* |
| Esholt (Emmerdale) | 750,000* |
| National Museum of Film, Photography and Television | 737,098 |
| Ilkley Moor | 550,000* |
| 1853 Gallery (Salt's Mill) | 500,000* |
| Bradford Cathedral | 202,200 |
| Bradford Industrial Museum | 170,000 |
| Keighley and Worth Valley Steam Railway | 141,028 |
| Brontë Parsonage | 101,900 |
| Cliffe Castle | 99,739 |
| Cartwright Hall | 90,477 |
| Bolling Hall | 37,908 |
| East Riddlesden Hall | 34,923 |
| Yorkshire Car Collection | 24,000 |

Source: Davidson and Maitland (1997).
* Estimated figure.

Edwardian theatres); and Salt's Mill, a converted woollen mill that contains shops, restaurants and a major art gallery showcasing the work of David Hockney (Table 8:3).

These attractions have been built into promotional strategies that focus upon short-break forms of tourism with strong thematic foci – 'In the Steps of the Brontës', 'Industrial Heritage' and (reflecting the distinctiveness of the Asian community in Bradford) 'Tastes of Asia' all being successful initiatives with clear heritage links. By 1994, estimated annual visitor levels had reached almost 5 million (see Figure 8:4), although a significant proportion of these would be day visitors rather than tourists making overnight stays. Indeed, one of the constraints on the development of tourism in Bradford that has emerged is a shortage of hotel beds, a matter that the city council is now addressing as the values of conference-based tourism have become more apparent. Overall, tourism is currently estimated to be worth £64 million to the local economy. As a different mark of success, popular destinations are already revealing the stresses and strains of congestion, pollution and disruption to local life that tourism can create. The small village of Esholt (which masquerades as *Emmerdale* in the TV soap) draws an estimated 500,000 visitors to a settlement without a single public lavatory!

**Figure 8:4** *Growth of visitor attendance at Bradford tourist attractions, 1986–94*

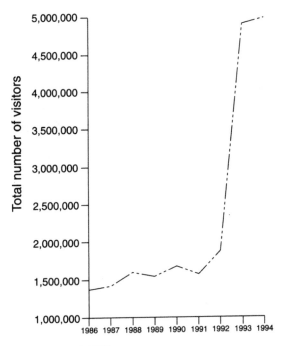

Source: Davidson and Maitland (1997).

The case of Bradford demonstrates several points. It shows the power of heritage attractions to promote new tourism destinations but, more importantly, it shows the importance of images of places. To succeed, Bradford had to reinvent itself, to cast off the old images of the industrial city and through conscious policies of investment and promotion, construct a new set of images that would appeal to the fickle and unpredictable tastes of the tourist. The fact that it was able to do so successfully and provide a model for similar places seeking to follow the same path tells us just how flexible the modern geographies of tourism are becoming.

Source: Davidson and Maitland (1997).

The composite effect of these changes has been to define new tourism geographies in countries such as Britain where heritage promotion forms a central plank in tourism development strategies. So alongside the established sites of heritage tourism – London, Canterbury, Oxford, Chester, York, Edinburgh – are a host of sites that were once the antithesis of tourism – a source of tourists rather than a destination – but places to which the visitors now flock. Unfashionable places such as Liverpool, Manchester, Wigan, Gateshead, Bradford, Stoke-on-Trent and Dudley now have a local tourist industry, based around heritage, which provides a valuable new medium for economic revival and a reassertion of a distinctive local sense of place. Box 8:1 outlines recent tourism development in one 'new' tourist city: Bradford.

## Theme parks

The development of theme parks as tourist attractions illustrates several aspects of the contemporary redefinition of tourism practices and places.

1 They are the quintessential post-modern spaces with their overt and conscious mixing of styles and deliberate confusion of the real with the artificial.
2 They represent the globalisation and homogenisation of tourism cultures as the parks have spread and multiplied from their origins in North American amusement parks to reach Britain, continental Europe and the Pacific Rim of Asia (Japan, Korea, Australia, etc.).
3 They appeal strongly to the 'post-tourists', the playful consumers of superficial signs and surfaces that some writers see as embodying the new age of tourism.
4 Theme park developments also illustrate very effectively the idea of invented places. This operates at two levels. First, in many parks the

visual and contextual fabric is often an invention since it portrays imaginary characters and places that circulate around cartoon characters, fairy stories, myths or legends. 'Magic Kingdoms' and 'Fantasy Lands' are popular constructions in theme parks across the world, whilst themes that have a stronger grounding in reality, such as Disney's 'Frontier Land' and 'Main Street USA', present idealised and selective recollections. Second, theme parks are quite capable of inventing new tourism geographies by the manner in which they are located. Whilst some ventures have gravitated towards established tourism areas, such as the theme park developments in Florida, others have been obliged (through their considerable land requirements) to take on greenfield sites in places where tourism is not currently present. The original Disney development at Anaheim, Los Angeles (opened in 1955), was located in a nondescript zone on the city fringe where the tourist stock amounted to just seven rather modest motels. Similarly, the subsequent development of Euro-Disney on a 2,000-ha site at Marne-la-Vallée, 32 km east of Paris, although close to a major tourism city, also introduced large-scale tourism to an area that had only been lightly affected previously.

The specific character of theme parks varies from place to place, and this has led to some problems in their definition. For purposes of this discussion a theme park is viewed as a self-contained family entertainment complex designed around landscapes, settings, rides, performances and exhibitions that reflect a common theme or set of themes. The leading exponent of the modern theme park, the Disney Corporation, reveals in its parks a remarkable synergy of resort development, state-of-the-art rides and amusements, as well as integral cross-references to the Corporation's own film and television products. But where Disney leads, others have followed, as the theme park concept has been replicated and reproduced in a growing number of settings across the globe.

Initially, the theme park grew out of fairground-style amusement parks of the 1920s, and a number of successful ventures had been established in North America and Europe prior to the opening of Disney's park at Anaheim. (Disney borrowed several ideas from the Efteling Park in the Netherlands, for example.) But the success of Disney's initial venture encouraged others to enter the field, including major entertainment corporations and, particularly, film companies. The successful development of American parks encouraged expansion in European theme parks from the late 1970s and East Asian and Australian parks also the case that whilst there has been some convergence in 'official' definitions (i.e. those used by tourism organisations, governments and international forums such as the United Nations (UN)), public perceptions of what constitutes a tourist and the activity of tourism may still differ quite markedly.

We may, however, tease out some basic technical definitions of tourists and tourism as a starting point. Dictionaries, for example, commonly explain a 'tourist' as a person undertaking a tour – a circular trip that is usually made for business, pleasure or education, at the end of which one returns to the starting point, normally the home. 'Tourism' is habitually viewed as a composite concept involving not just the temporary movement of people to destinations that are removed from their normal place of residence but, in addition, the organisation and conduct of their activities and of the facilities and services that are necessary for meeting their needs.

Simple statements of this character are actually quite effective in drawing attention to the core elements that distinguish tourism as an area of activity:

- They give primacy to the notion that tourism necessarily involves travel but that the relocation of people is a temporary one.
- They make explicit the idea that motivations for tourism may come from one (or more than one) of a variety of sources. We tend to think of tourism as being associated with pleasure motives, but it can also embrace business, education, health or religion as a basis for travelling.
- They draw attention to the fact that the activity of tourism requires an accessible supporting infrastructure of transport, accommodation, marketing systems, entertainment and attractions that together form the basis for the tourism industries.

The spirit of these conceptions of tourism is, however, only implicit in the WTO definition published in 1991. This took a rather general view of tourism as:

> 'the activities of a person travelling to a place outside his or her usual environment for less than a specified period of time and whose main purpose of travel is other than the exercise of an activity remunerated from within the place visited'.

A number of writers have suggested that this definition needs further qualification by recognising that the time frame should normally be more than one day (thereby involving an overnight stop – a distinguishing feature that has been central to many attempts to define tourism), but no more than one year. However, neither the WTO definition, nor an earlier statement from the International Union of Tourism Organisations (IUOTO), which saw tourists as 'any person visiting a country, region or place other than that in which he or she has their usual place of residence', necessarily places an emphasis upon overnight stops as a defining feature of tourism. This view finds favour with a growing number of authors who argue that the actions of day visitors and excursionists are often indistinguishable in cause and effect from those of staying visitors and that these short-term visitors should also be considered as tourists.

This raises the wider issue of the relationship between tourism, recreation and leisure. As areas of academic study (and not least within the discipline of geography), a tradition of separate modes of investigation has emerged within these three fields, with particular emphasis upon the separation of tourism. Unfortunately, the terms 'leisure' and 'recreation' are themselves contested, but if we view 'leisure' as being related either to free time and/or to a frame of mind in which the individual believes themself to be 'at leisure' and 'recreation' as being activity or experience set within the context of leisure, then tourism (as defined so far) is clearly congruent with major areas of recreation and leisure. Not only does a significant portion of tourism activity take place in the leisure time/space framework, but it also centres upon recreation activities and experiences (for example, sightseeing, travelling for pleasure, leisure shopping) that may occur with equal ease within leisurely contexts that exist outside the framework of tourism. Convergence in the experience of leisure, recreation and tourism is also reinforced by the manner in which tourism is increasingly permeating day-to-day leisure lifestyles. We read about tourism in newspapers or magazines and view television travel shows; we spend leisure time reviewing home videos or photo albums of previous trips and actively planning future ones; and we import experiences of travel into our home lives, for example by eating at foreign restaurants, practising our winter sports at the local dry slope, visiting the leisure centre to acquire an artificial tan before the Mediterranean holiday, or by including foreign clothing styles within our wardrobe.

In approaching the study of tourism, therefore, we need to understand that the relationships between leisure, recreation and tourism are much closer than the disparate manner in which they are treated in textbooks might suggest. There is considerable common ground in the major motivations for participation (attractions of destinations, events and experiences; social contacts; exploration), in the factors that facilitate engagement with activity (discretionary income, mobility, knowledge of opportunity) and the rewards (pleasure, experience, knowledge or memories) that we gain from tourism, recreation and leisure. Figure 1:1 represents this relationship diagrammatically and draws attention both to those areas of tourism that coincide with the realms of leisure and recreation and those which lie outside or where linkages are less clear. Even here we must be careful, however, for the business tourist (for example) will almost certainly spend some of the time during their trip engaging in recreational or leisurely pursuits. It may be more helpful, therefore, to visualise Figure 1:1 as differentiating *forms of experience* rather than *categories of visitor* and imagining some individuals moving between the overlapping spheres, even within the context of a single trip.

**Figure 1:1** *Relationship between leisure, recreation and tourism*

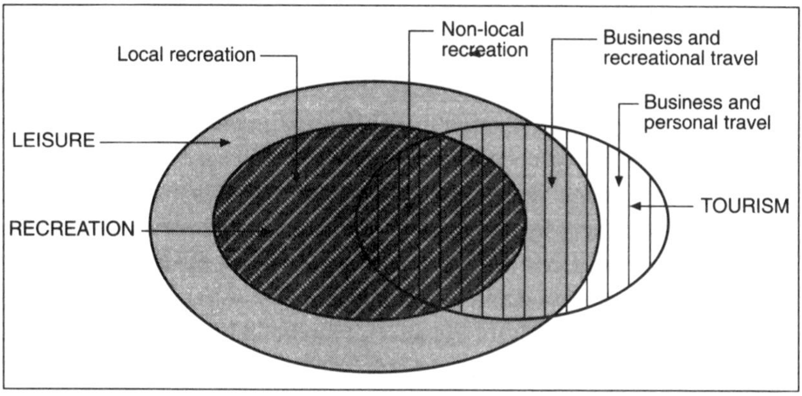

Source: Adapted from Murphy (1985).

## Problems in the study of tourism

The definitional complexities of tourism and the uncertain linkages with the allied fields of recreation and leisure are basic problems that confront the student of tourism, though they are not the sole difficulties. Four further problems merit brief attention at this introductory stage.

First, in later chapters I shall use a range of statistics to map out the basic dimensions and patterns of tourism, but it is important to appreciate that in many situations, comparability across space and time is made difficult or sometimes impossible by variation in official practice in distinguishing and recording the levels of tourist activity. Some countries do not even count the arrivals of foreign nationals at their borders. Since July 1995 relaxation of border controls between member states of the EU that are signatories to the Schengen Treaty (Belgium, the Netherlands, Luxembourg, France, Germany, Portugal and Spain) now permits largely unrestricted (and hence undocumented) movement of tourists between these countries. Elsewhere, the presence of foreign nationals may be recorded at points of entry, although local definitions of tourist status or failure to identify precise motives for visiting can also lead to an inability to enumerate tourists exactly. Some states count business travellers as tourists whilst others may not. Within states, tourism statistics may also be compiled through sample surveys of visitors or by reference to hotel registrations, although these will naturally be selective and prone to imprecision. Hotel-based figures, for example, will overlook those visitors who lodge with friends or relatives. Data, therefore, are seldom directly comparable and always need to be treated with some caution.

Second, there are problems inherent in the definition of tourism as an industry, even though there are clear practical advantages in delineating tourism as a coherent and bounded area of activity. It has been argued that designating tourism as an 'industry' establishes a framework within which activity and associated impacts may be measured and recorded, and, more critically, provides a form of legitimisation for an activity that has often struggled to gain the strategic recognition of political and economic analysts. However, tourism in practice is a nebulous area and the notion that it may be conceived as a distinctive industry with a definable product and measurable flows of associated goods, labour and capital has in itself been a problem. Conventionally, an industry is defined as a group of firms engaged in the manufacture or production of a given product or service. In tourism, though, there are many products and services, some tangible (provision of accommodation, entertainment and the production of gifts and souvenirs), others less so (creation of experience, memories or social contact). Many of the firms that service tourists also provide the same service to local people who do not fall into the category of tourists, however it may be cast. Tourism is not, therefore, an industry in any conventional sense. It is really a collection of industries which experience varying levels of dependence upon visitors, a dependence that alters through both space and time.

## Box 8:2

### Recent growth of theme parks: the case of Japan

The development of theme parks in Japan illustrates not only the rapidity with which these tourist spaces have become popular, but also the importance of thematic diversity, location and accessibility, effective management and political support in securing the establishment of successful parks.

The first recognised theme park – a reconstruction of a Meiji village – opened in 1965 at Inuyama City, some 200 km west of Tokyo. This was followed in 1975 by a second park at Kyoto that is based on a theme of films. Until 1983, these two parks were Japan's only examples, but between 1983 and 1991 a further twenty-five parks were opened (see Figure 8:5).

**Figure 8:5** *Development of theme parks in Japan*

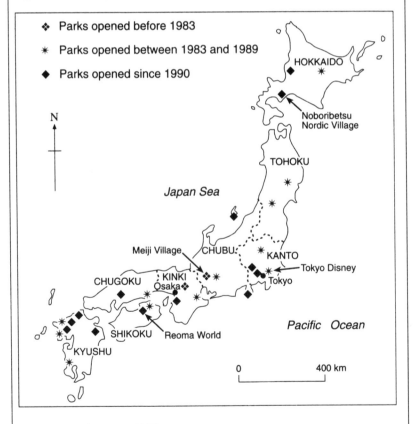

Source: Adapted from Jones (1994).

The catalyst to this remarkable expansion was the opening in 1983 of Tokyo Disneyland at Urayasu City. Set in the heart of Japan's most populated area, with 30 million people resident within a 50-km radius, the location and the high levels of accessibility to Japan's major cities have been key variables in the dramatic success of this venture. The advantages of location have been further reinforced by the skills of the Japanese as business managers and their instinctive attention to the efficiency of customer services, the cleanliness of the park and the smoothness of its operations. Tokyo Disney also demonstrates the importance of ongoing investment and the production of new rides and attractions as a means of retaining their market share – a lesson that smaller ventures have been slower to learn.

The success of Tokyo Disney spawned a number of direct imitations but also encouraged the promotion of parks centred on other themes. Japanese parks mirror the cultures in which they are produced, so alongside the themes of fantasy and adventure (which seem to possess a universal appeal) are parks that reflect Japanese interest in their history, nature, technology and – rather unusually – the cultures of other countries. The latter theme (which includes parks centred on reconstructions of Nordic villages, German towns of the Middle Ages, New Zealand farms and Dutch cities) produces decontextualised spaces that often leave foreign visitors who know the originals quite bemused, but which Japanese tourists apparently take in their stride.

Part of the growth and diversification of Japanese theme parks is explained by the natural enthusiasm of the urban Japanese for tourism and travel, but it is also the product of political support from the Japanese government and its prefectures (local authorities), together with substantial investment from enterprises outside tourism. Tokyo Disney, for example, is owned by real estate and railway companies, with local government as a minor shareholder. Several of the less well positioned parks (such as Noboribetsu Nordic Village on Hokkaido and Reoma World Water Park on the island of Shikoko) would not have come into being without strong local political support and investment. In Japan, the theme park is a mark of local success and for some communities is literally a means of getting onto the map.

However, it is becoming clear that there are limits to the growth of theme parks in Japan. Whilst the 1990s have seen several more large projects added to the list, concerns over problems of identity within the market, the spiralling costs of investment and reinvestment, as well as uncertainties over the longer-term impacts of theme park development, have led to cancellations and postponements of other planned ventures.

Source: Jones (1994).

during the 1980s. The area of fastest growth in theme parks is now centred on the Pacific Rim, and Box 8:2 illustrates the trend with an outline of a case study of theme park development in Japan. Alongside the rapid expansion of theme parks, other tourist attractions have also incorporated themed areas as part of their planned developments.

Shopping malls such as West Edmonton (Alberta, Canada) and The Mall of the Americas at Minneapolis (USA) are spectacular examples of new tourist spaces in which theming is a prominent aspect of the appeal.

The spatial expansion of theme parks as tourist attractions is, of course, a reflection of the success of the concept and its almost universal appeal – something that is strongly reflected in their capacity to draw huge numbers of visitors. These are family attractions that, perhaps surprisingly, also appeal to older tourists. This is especially true when the theme park focuses upon historical, natural or cultural attractions, rather than 'white-knuckle' rides alone. In Britain, Alton Towers and Blackpool's Pleasure Beach consistently dominate the rankings of attractions at which visitors pay for entry, with Alton Towers drawing around 2.5 million visitors in a typical recent year. This pales almost into insignificance, however, compared to the market leaders in North America and the Far East. Tokyo Disney attracted almost 16 million visitors in 1993 while Disney's original venture at Anaheim drew 11.4 million people in the same year. In fact, the combined total attendance at the four Disney parks in the USA (Disneyland, Magic Kingdom, EPCOT and Disney World/MGM – the last three all in Florida) reached a staggering 41 million people. Aggregate attendance at theme parks in the USA and Canada in 1990 touched 160 million, an increase of 24 per cent over the decade since 1980 (Figure 8:6).

**Figure 8:6   *Growth of theme park attendance in Canada and the USA, 1980–94***

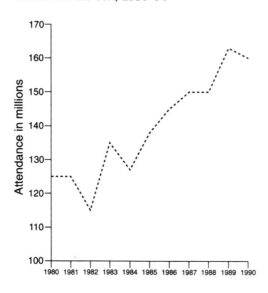

Source: Adapted from Loverseed (1994).

The spatial distribution of the major parks is interesting. As the Japanese case shows, there are clear advantages to being close to major urban markets and/or established tourism regions. In the USA, as Figure 8:7 shows, the largest parks generally favour the warmer states such as Florida and

**Figure 8:7 Distribution of major theme parks in the USA**

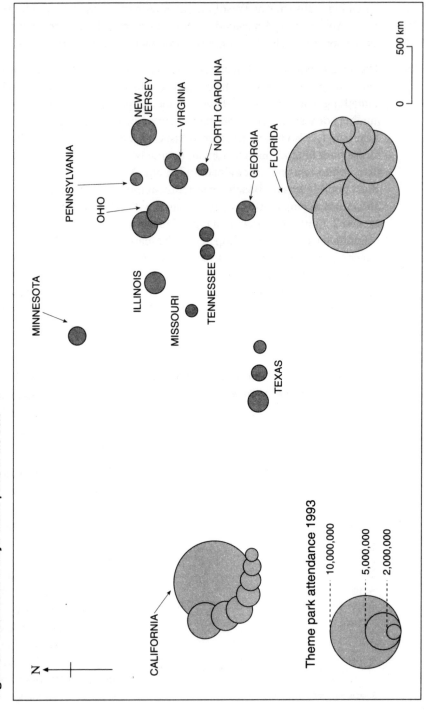

N

MINNESOTA

CALIFORNIA

ILLINOIS

MISSOURI

TEXAS

TENNESSEE

PENNSYLVANIA

OHIO

NEW JERSEY

VIRGINIA

NORTH CAROLINA

GEORGIA

FLORIDA

Theme park attendance 1993

10,000,000

5,000,000

2,000,000

0        500 km

Source: Adapted from Loverseed (1994).

California, since these represent the preferred destinations for American tourists in general. The more attractive climates in these locations clearly favour outdoor parks. (This was a requirement that Disney discovered to its cost in the near-disastrous opening of Euro-Disney in the damp and often cold outskirts of Paris.) But interestingly, parks can also be developed successfully in less propitious locations. Some of the most rapid rates of expansion in visits to American theme parks have been recorded in less popular areas such as Illinois, Ohio, Missouri, Tennessee and Kentucky. Similarly, in Britain, the most popular theme park, Alton Towers, is buried deep in the lanes of rural north-east Staffordshire, one of the least visited counties in England – especially if visitors to the park are deducted from aggregate totals! The capacity of tourism to invent new places is indeed remarkable.

## Conclusion

The case studies and examples that have been employed in this final chapter demonstrate particularly well how the dynamic quality of contemporary tourism has become one of its defining features. The traditional patterns of annual holidays by the sea in high summer, which in some locations endured for the best part of a century, are fast receding into individual and collective memories, replaced by new tourism geographies that emphasise different ways of seeing the world we inhabit. The seasides, lakes and mountains that supported the development of tourism for a large part of the twentieth century still attract, but so, increasingly, do shopping malls, centres of culture and heritage, sporting venues, theme parks, centres of industry, rainforests, savannahs, wildernesses and icefields. The (post-)modern tourist, no longer confined to traditional times and spaces, has become virtually ubiquitous. So, too, have tourism's effects, impacts and influences.

For human geographers, a central theme within the discipline is interpreting and understanding our changing world – a world in which geographic patterns are constantly being reworked by powerful forces of change: population shifts; new patterns of economic production and consumption; evolving social and political structures; new forms of urbanism; and globalisation and the compressions of time and space that are the product of the ongoing revolutions in information technology and telecommunications. This book has attempted to show how tourism has also come to be a major force for change – an integral and indispensable part of the places in which we live, their economies and their societies. When scarcely a corner of the globe remains untouched by the influence of tourism, this is a phenomenon that we can no longer ignore.

## Summary

One of the defining features of tourism is its fluidity in time and space and, in this final chapter, we have explored some of the ways in which new tourist geographies and associated places are formed. Whilst many factors contribute to the formation of new patterns of tourism, social and culturally constructed images of destinations are seen as especially important. The shifting focus of the tourist gaze and the cultural filters that mediate our individual and collective views of tourist places directly and indirectly shape the spatial patterns of tourism at an increasingly globalised scale, producing evolving patterns in both the production and consumption of tourist places.

## Discussion questions

1  Does it matter if tourist experiences of places are 'unreal'?
2  How far would you agree that tourist practices represent an essentially visual consumption of places?
3  In what ways do holiday brochures present selective representations of tourism places?
4  How significant is the notion of 'difference' in the promotion of tourist destinations?
5  What are the defining features of the 'new' forms of tourism and how will such forms lead to the production of new tourist geographies?

## Further reading

The concept of the tourist gaze and an interesting collection of essays on cultural inventions of place are to be found in:
Urry, J. (1990) *The Tourist Gaze*, London: Sage.
—— (1995) *Consuming Places*, London: Routledge.

A varied set of recent essays on image formation in tourism may be found in:
Selwyn, T. (ed.) *The Tourist Image: Myths and Myth Making in Tourism*, Chichester: John Wiley.

whilst marketing approaches to image formation are considered in:

Ashworth, G.J. (1991) 'Products, places and promotion: destination images in the analysis of the tourism industry'. In Sinclair, M.T. and Stabler, M.J. (eds) *The Tourism Industry: An International Analysis*, Wallingford: CAB International: 121–142.

Issues of authenticity and the nature of attractions are discussed by:

Cohen, E. (1994) 'Contemporary tourism – trends and challenges: sustained authenticity or contrived post-modernity?' In Butler, R. and Pearce, D. (eds) *Change in Tourism*, London: Routledge: 12–29.

The recent growth of cultural and heritage tourism is discussed in:

Prentice, R. (1993) *Tourism and Heritage Attractions*, London: Routledge.

Richards, G. (ed.) (1996) *Cultural Tourism in Europe*, Wallingford: CAB International.

For a critical evaluation of the work of Walt Disney see:

Bryman, A. (1993) *Disney and His Worlds*, London: Routledge.

whilst examples of recent analyses of theme park developments may be found in:

Jones, T.S.M. (1994) 'Theme parks in Japan'. In Cooper, C.P. and Lockwood, A. (eds) *Progress in Tourism, Recreation and Hospitality Management*, Vol. 6, Chichester: John Wiley: 111–125.

Loverseed, H. (1994) 'Theme parks in North America', *Travel and Tourism Analyst*, No. 4: 51–63.

For a wider discussion of alternative forms of tourism see:

Smith, V.L. and Eadington, W.R. (eds) (1994) *Tourism Alternatives: Potentials and Problems in the Development of Tourism*, Chichester: John Wiley.

# Bibliography

Albert-Pinole, I. (1993) 'Tourism in Spain'. In Pompl, W. and Lavery, P. (eds) (1993): 242–261.

Alipour, H. (1996) 'Tourism development within planning paradigms: the case of Turkey', *Tourism Management*, Vol. 17 No. 5: 367–377.

Ap, J. (1992) 'Residents' perceptions on tourism impacts', *Annals of Tourism Research*, Vol. 19 No. 4: 665–690.

Archer, B.H. (1989) 'Tourism and island economies: impact analyses'. In Cooper, C.P. (ed.) (1989): 125–134.

Ashworth, G.J. (1991) 'Products, places and promotion: destination images in the analysis of the tourism industry'. In Sinclair, M.T. and Stabler, M.J. (eds) (1991): 121–142.

Ashworth, G.J. and Dietvorst, A.G.J. (eds) (1995) *Tourism and Spatial Transformations*, Wallingford: CAB International.

Baum, T. (1994) 'The development and implementation of national tourism policies', *Tourism Management*, Vol. 15 No. 3: 185–192.

Becheri, E. (1991) 'Rimini and Co – the end of a legend?', *Tourism Management*, Vol. 12 No. 3: 229–235.

Blundell, V. (1993) 'Aboriginal empowerment and souvenir trade in Canada', *Annals of Tourism Research*, Vol. 20 No. 1: 64–87.

Boniface, B.G. and Cooper, C.P. (1994) *The Geography of Travel and Tourism*, Oxford: Butterworth Heinemann.

Briguglio, L., Butler, R.W., Harrison, D. and Filho, W.L. (1996) *Sustainable Tourism in Islands and Small States: Case Studies*, London: Pinter.

British Tourist Authority (1995) *Digest of Tourist Statistics No. 18*, London: BTA.

—— (1997) *English Heritage Monitor 1997*, London: BTA.

British Travel Association (1969) *Patterns in British Holiday-making 1951–1968*, London: BTA.

Brown, D. (1996) 'Genuine fakes'. In Selwyn, T. (ed.) (1996): 33–47.

Burton, R.C.J. (1994) 'Geographical patterns of tourism in Europe'. In Cooper, C.P. and Lockwood, A. (eds) (1994): 3–25.

Butler, R.W. (1980) 'The concept of a tourist area cycle of evolution:

implications for management of resources', *Canadian Geographer*, Vol. 24 No. 1: 5–12.
—— (1993) 'Tourism development in small islands: past influences and future directions'. In Lockhart, D. *et al.* (eds) (1993): 71–91.
—— (1993) 'Pre- and post-impact assessment of tourism development'. In Pearce, D.G. and Butler, R.W. (eds) (1993): 135–155.
—— (1994) 'Alternative tourism: the thin end of the wedge'. In Smith, V.L. and Eadington, W.R. (eds) (1994): 31–46.
Butler, R.W. and Pearce, D.G. (eds) (1994) *Change in Tourism: People, Places, Processes*, London: Routledge.
Choy, D.J.L. (1991) 'National tourism planning in the Philippines', *Tourism Management*, Vol. 12 No. 3: 245–251.
—— (1993) 'Alternative roles of national tourism organizations', *Tourism Management*, Vol. 14 No. 5: 357–365.
—— (1995) 'The quality of tourism employment', *Tourism Management*, Vol. 16 No. 2: 129–137.
Cohen, E. (1972) 'Towards a sociology of international tourism', *Social Research*, Vol. 39: 164–182.
—— (1979) 'Rethinking the sociology of tourism', *Annals of Tourism Research*, Vol. 6 No. 1: 18–35.
—— (1993) 'Open-ended prostitution as a skilful game of luck: opportunity, risk and security among tourist-oriented prostitutes in a Bangkok soi'. In Hitchcock, M. *et al.* (eds) (1993): 155–178.
—— (1994) 'Contemporary tourism – trends and challenges: sustainable authenticity or contrived post-modernity?' In Butler, R. and Pearce, D. (eds) (1994): 12–29.
Cooper, C.P. (ed.) (1989) *Progress in Tourism, Recreation and Hospitality Management*, Vol. 1, London: Belhaven.
—— (ed.) (1990) *Progress in Tourism, Recreation and Hospitality Management*, Vol. 2, London: Belhaven.
—— (1990) 'Resorts in decline: the management response', *Tourism Management*, Vol. 11 No. 1: 63–67.
—— (1995) 'Strategic planning for sustainable tourism: the case of offshore islands of the UK', *Journal of Sustainable Tourism*, Vol. 3 No. 4: 191–207.
Cooper, C.P. and Lockwood, A. (eds) (1994) *Progress in Tourism, Recreation and Hospitality Management*, Vol. 6, Chichester: John Wiley.
Cooper, C.P. and Ozdil, I. (1992) 'From mass to "responsible" tourism: the Turkish experience', *Tourism Management*, Vol. 13 No. 4: 377–386.
Corbin, A. (1995) *The Lure of the Sea*, London: Penguin.
Craig-Smith, S.J. and Fagence, M. (1994) 'A critique of tourism planning in the Pacific'. In Cooper, C.P. and Lockwood, A. (eds) (1994): 92–110.
Craik, J. (1991) *Resorting to Tourism: Cultural Policies for Tourist Development in Australia*, Sydney: Allen and Unwin.

—— (1995) 'Are there cultural limits to tourism?', *Journal of Sustainable Tourism*, Vol. 3 No. 2: 87–98.

Craik, W. (1994) 'The economics of managing fisheries and tourism in the Great Barrier Reef Marine Park'. In Munasinghe, M. and McNeely, J. (eds) (1994): 339–348.

Dann, G. (1996) 'The people of tourist brochures'. In Selwyn, T. (ed.) (1996): 61–81.

Davidson, R. (1992) *Tourism in Europe*, London: Pitman.

Davidson, R. and Maitland, R. (1997) *Tourism Destinations*, London: Hodder and Stoughton.

Davidson, T.L. (1994) 'What are travel and tourism: are they really an industry?' In Theobald, W. (ed.) (1994): 20–26.

Dieke, P.U.C. (1994) 'The political economy of tourism in the Gambia', *Review of African Political Economy*, No. 62: 611–627.

Eadington, W.R. (1994) 'The emergence of casino gaming as a major factor in tourism markets: policy issues and considerations'. In Butler, R. and Pearce, D. (eds) (1994): 159–186.

Economics Intelligence Unit (1993) 'Turkey', *International Tourism Report* No. 3: 77–97.

—— (1995) 'Thailand', *International Tourism Report* No. 3: 67–81.

Foley, M. (1996) 'Cultural tourism in the United Kingdom'. In Richards, G. (ed.) (1996): 283–309.

Freitag, T.G. (1994) 'Enclave tourism development: for whom the benefits roll?', *Annals of Tourism Research*, Vol. 21 No. 3: 538–554.

Gant, R. and Smith, J. (1992) 'Tourism and national development planning in Tunisia', *Tourism Management*, Vol. 13 No. 3: 331–336.

Getz, D. (1994) 'Residents' attitudes towards tourism: a longitudinal study in Spey Valley, Scotland', *Tourism Management*, Vol. 15 No. 4: 247–258.

Gilbert, D.C. (1990) 'Conceptual issues in the meaning of tourism'. In Cooper, C.P. (ed.) (1990): 4–27.

Gilbert, E.M. (1975) *Brighton: Old Ocean's Bauble*, Hassocks: Flare Books. (Originally published by Methuen, London, 1954.)

Gillmor, D.A. (1996) 'Evolving air-charter tourism patterns: change in outbound traffic from the Republic of Ireland', *Tourism Management*, Vol. 17 No. 1: 9–16.

Gomez, M.J.M. (1995) 'New tourism trends and the future of Mediterranean Europe', *Tijdschrift voor Economische en Sociale Geografie*, Vol. 86 No. 1: 21–31.

Gordon, I. and Goodall, B. (1992) 'Resort cycles and development processes', *Built Environment*, Vol. 18 No. 1: 41–55.

Graburn, N.H.H. (1983) 'The anthropology of tourism', *Annals of Tourism Research*, Vol. 10 No. 1: 9–33.

—— (1994) 'The past in the present in Japan: nostalgia and neo-traditionalism in contemporary Japanese domestic tourism'. In Butler, R. and Pearce, D. (eds) (1994): 47–70.

Grahn, P. (1991) 'Using tourism to protect existing culture: a project in Swedish Lapland', *Leisure Studies*, Vol. 10 No. 1: 33–47.

Gratton, C. and Van der Straaten, J. (1994) 'The environmental impact of tourism in Europe'. In Cooper, C.P. and Lockwood, A. (eds) (1994): 147–161.

Gunn, C.A. (1988) *Tourism Planning*, New York: Taylor and Francis.

Hall, C.M. (1994) 'Sex tourism in South-east Asia'. In Harrison, D. (ed.) (1994): 64–74.

Hall, D.R. (1992) 'The challenge of international tourism in eastern Europe', *Tourism Management*, Vol. 13 No. 1: 41–44.

Harrison, D. (ed.) (1994) *Tourism and the Less Developed Countries*, London: Belhaven/John Wiley.

Hawkins, R. and Royden, A. (1991) 'Tourism development in coastal resorts of Europe', *Journal of Regional and Local Studies*, Vol. 11 Pts 1–2: 33–41.

Heng, T.M. and Low, L. (1990) 'Economic impact of tourism in Singapore', *Annals of Tourism Research*, Vol. 17: 246–269.

Hitchcock, M., King, V.T. and Parnwell, M.J.G. (eds) (1993) *Tourism in South-East Asia*, London: Routledge.

Hunter, C. and Green, H. (1995) *Tourism and the Environment: A Sustainable Relationship?*, London: Routledge.

Inskeep, E. (1991) *Tourism Planning: An Integrated and Sustainable Development Approach*, New York: Van Nostrand Reinhold.

Ironbridge Gorge Museum Trust (1996) *Tourist Survey 1995*, Ironbridge: The Trust.

Iso-Ahola, S.E. (1982) 'Towards a social psychological theory of tourism motivation', *Annals of Tourism Research*, Vol. 9 No. 2: 256–262.

Jansen-Verbeke, M. and van de Weil, E. (1995) 'Tourism planning in urban revitalization projects: lessons from the Amsterdam Waterfront development'. In Ashworth, G.J. and Dietvorst, A.G.J. (eds) (1995): 129–145.

Johnson, J.D., Snepenger, D.J. and Akis, S. (1994) 'Residents' perceptions of tourism development', *Annals of Tourism Research*, Vol. 21 No. 3: 629–642.

Johnson, M. (1995) 'Czech and Slovak tourism: patterns, problems and prospects', *Tourism Management*, Vol. 16 No. 1: 21–28.

Jones, T.S.M. (1994) 'Theme parks in Japan'. In Cooper, C.P. and Lockwood, A. (eds) (1994): 111–125.

Kaspar, C. and Laesser, C. (1993) 'Tourism in non-EC countries: the case of Switzerland and Austria'. In Pompl, W. and Lavery, P. (eds) (1993): 324–340.

Khunaphante, P. (1992) 'Thailand: the preferable destination for the European tourist'. In *Tourism in Europe – the 1992 Conference*, Houghton-le-Spring, Centre for Travel and Tourism Research: K1–K13.

King, V.T. (1993) 'Tourism and culture in Malaya'. In Hitchcock, M. *et al.* (eds) (1993): 99–116.

Krippendork, J. (1987) *The Holiday Makers*, Oxford: Butterworth Heinemann.

Lane, P. (1992) 'The regeneration of small to medium sized seaside resorts'. In *Tourism in Europe – the 1992 Conference*, Houghton-le-Spring, Centre for Travel and Tourism Research: L1–L13.

Lanfant, M.-F. (1995) 'Tourism, internationalization and identity'. In Lanfant, M.-F. *et al.* (eds) (1995): 24–43.

Lanfant, M.-F., Allcock, J.B. and Bruner, E.M. (eds) (1995) *International Tourism: Identity and Change*, London: Sage.

Latham, J. (1994) 'International tourism statistics'. In Cooper, C.P. and Lockwood, A. (eds) (1994): 327–333.

Lavery, P. (1993) 'Tourism in the United Kingdom'. In Pompl, W. and Lavery, P. (eds) (1993): 129–148.

Law, C.M. (1992) 'Urban tourism and its contribution to economic regeneration', *Urban Studies*, Vol. 29 Nos. 3/4: 599–618.

Lawrence, K. (1994) 'Sustainable tourism development'. In Munasinghe, M. and McNeely, J. (eds) (1994): 263–272.

Lea, J. (1988) *Tourism and Development in the Third World*, London: Routledge.

Lee, G.P. (1987) 'Tourism as a factor in development co-operation', *Tourism Management*, Vol. 8 No. 1: 2–19.

Lockhart, D. (1993) 'Tourism to Fiji: Crumbs off a rich man's table?', *Geography*, Vol. 78 No. 3: 318–323.

Lockhart, D., Drakakis-Smith, D. and Schembri, J. (eds) (1993) *The Development Process in Small Island States*, London: Routledge.

Loverseed, H. (1994) 'Theme parks in North America', *Travel and Tourism Analyst*, No. 4: 51–63.

Lundgren, J.O.L. (1992) 'Transport infrastructure development and tourist travel: case Europe'. In *Tourism in Europe – the 1992 Conference*, Houghton-le-Spring, Centre for Travel and Tourism Research: L25–L36.

MacCannell, D. (1992) *Empty Meeting Grounds: The Tourist Papers*, London: Routledge.

Mathieson, A. and Wall, G. (1982) *Tourism: Economic, Physical and Social Impacts*, Harlow: Longman.

—— (1997) *Tourism: Change, Impacts and Opportunities*, Harlow: Longman.

Meethan, K. (1996) 'Place, image and power: Brighton as a resort'. In Selwyn, T. (ed.) (1996): 179–196.

Mercer, D. (1994) 'Native peoples and tourism: conflict and compromise'. In Theobald, W. (ed.) (1994): 124–145.

Milne, S. (1992) 'Tourism and development in South Pacific microstates', *Annals of Tourism Research*, Vol. 19 No. 2: 191–212.

Mings, R. and Chulikpongse, S. (1994) 'Tourism in far southern Thailand: a geographical perspective', *Tourism Recreation Research*, Vol. 19 No. 1: 25–31.

Mitchell, L.S. and Murphy, P.E. (1991) 'Geography and tourism', *Annals of Tourism Research*, Vol. 18 No. 1: 57–70.

Moore, K., Cushman, G. and Simmons, D. (1995) 'Behavioural conceptualizations of tourism and leisure', *Annals of Tourism Research*, Vol. 22 No. 1: 67–75.

Munasinghe, M. and McNeely, J. (eds) (1994) *Protected Areas Economics and Policy: Linking Conservation and Sustainable Development*, Washington, DC: World Bank.

Murphy, P.E. (1985) *Tourism: A Community Approach*, London: Routledge.

—— (1994) 'Tourism and sustainable development'. In Theobald, W. (ed.) (1994): 274–290.

North York Moors National Park (NYMNP) (1990) *Visitors and the National Park Landscape*, Danby: NYMNP Education Service.

—— (1993) *North York Moors National Park Visitor Survey 1991*, Helmsley: NYMNP.

O'Brien, K. (1990) *The UK Tourism and Leisure Market: Travel Trends and Spending Patterns*, London: The Economics Intelligence Unit, Special Report No. 2010.

Oppermann, M. (1992) 'International tourism and regional development in Malaysia', *Tijdschrift voor Economische en Sociale Geografie*, Vol. 83 No. 3: 226–233.

Organisation for Economic Co-operation and Development (1996) *Tourism Policy and International Tourism in OECD Countries 1993–1994*, Paris: OECD.

Page, S. (1994) *Transport for Tourism*, London: Routledge.

Page, S. and Thorn, K.J. (1997) 'Towards sustainable tourism planning in New Zealand: public sector planning responses', *Journal of Sustainable Tourism*, Vol. 5 No. 1: 59–75.

Parnwell, M.J.G. (1993) 'Environmental issues and tourism in Thailand'. In Hitchcock, M., King, V.T. and Parnwell, M.J.G. (eds) (1993): 286–302.

Patronato Provincial de Turismo de la Costa del Sol (1988) *La oferta y demanda turística de la Costa del Sol*, Torremolinos: PPTCS.

Pearce, D.G. (1987) *Tourism Today: A Geographical Analysis*, Harlow: Longman.

—— (1989) *Tourism Development*, Harlow: Longman.

—— (1994) 'Planning for tourism in the 1990s: an integrated, dynamic, multiscale approach'. In Butler, R.W. and Pearce, D.G. (eds) (1994): 229–244.

—— (1994) 'Alternative tourism: concepts, classifications and questions'. In Smith, V.L. and Eadington, W.R. (eds) (1994): 15–30.

Pearce, D.G. and Butler, R.W. (eds) (1993) *Tourism Research: Critiques and Challenges*, London: Routledge.

Pearce, P.L. (1993) 'Fundamentals of tourist motivation'. In Pearce, D.G. and Butler, R.W. (eds) (1993): 113–134.

—— (1994) 'Tourist–resident impacts: examples, explanations and emerging solutions'. In Theobald, W. (ed.) (1994): 103–123.

Perkins, H. (1971) *The Age of the Railway*, Newton Abbot: David and Charles.

Picard, M. (1993) 'Cultural tourism in Bali: national integration and regional differentiation'. In Hitchcock, M. *et al.* (eds) (1993): 71–98.

—— (1995) 'Cultural tourism in Bali'. In Lanfant, M.-F. *et al.* (eds) (1995): 44–66.

Pigram, J.J. (1993) 'Planning for tourism in rural areas: bridging the policy implementation gap'. In Pearce, D.G. and Butler, R.W. (eds) (1993): 156–174.

—— (1994) 'Resource constraints on tourism: water resources and sustainability'. In Butler, R. and Pearce, D. (eds) (1994): 208–228.

Pimlott, J.A.R. (1947) *The Englishman's Holiday: A Social History*, London: Faber.

Pitchford, S. R. (1995) 'Ethnic tourism and nationalism in Wales', *Annals of Tourism Research*, Vol. 22 No. 1: 35–52.

Plog, S. (1974) 'Why destination areas rise and fall in popularity', *Cornell Hotel Restaurant and Administration Quarterly*: 55–58.

Poirier, R.A. (1995) 'Tourism and development in Tunisia', *Annals of Tourism Research*, Vol. 22 No. 1: 157–171.

Pollard, J. and Rodriguez, R.D. (1993) 'Tourism and Torremolinos: recession or reaction to environment?', *Tourism Management*, Vol. 14 No. 4: 247–258.

Pompl, W. (1993) 'The liberalisation of European transport markets'. In Pompl, W. and Lavery, P. (eds) (1993): 55–79.

Pompl, W. and Lavery, P. (eds) (1993) *Tourism in Europe: Structures and Developments*, Wallingford: CAB International.

Prentice, R. C. (1993) *Tourism and Heritage Attractions*, London: Routledge.

—— (1994) 'Heritage: a key sector of the "new" tourism'. In Cooper, C.P. and Lockwood, A. (eds) (1994): 308–324.

Price, M.F. (1992) 'Patterns of development of tourism in mountain environments', *Geo Journal*, Vol. 27 No. 1: 87–96.

Prunier, E.K., Sweeney, A.E. and Geen, A.G. (1993) 'Tourism and the environment: the case of Zakynthos', *Tourism Management*, Vol. 14 No. 2: 137–141.

Przeclawski, K. (1993) 'Tourism as the subject of inter-disciplinary research'. In Pearce, D.G. and Butler, R.W. (eds) (1993): 9–19.

Richards, G. (ed.) (1996) *Cultural Tourism in Europe*, Wallingford: CAB International.

Romeril, M. (1989) 'Tourism – the environmental dimension'. In Cooper, C.P. (ed.) (1989): 102–113.

Ryan, C. (1991) *Recreational Tourism: A Social Science Perspective*, London: Routledge.

Scarborough Borough Council (1994) *Scarborough Borough Local Plan Fact Sheet – Tourism*, Scarborough: The Council.

Selwyn, T. (1996) *The Tourist Image: Myths and Myth Making in Tourism*, Chichester: John Wiley.

Shaw, G. and Williams, A.M. (1991) 'From bathing hut to theme park: tourism development in south west England', *Journal of Regional and Local Studies*, Vol. 11 No. 1/2: 16–32.

—— (1994) *Critical Issues in Tourism: A Geographical Perspective*, Oxford: Blackwell.

Sidaway, R. (1995) 'Managing the impacts of recreation by agreeing the Limits of Acceptable Change'. In Ashworth, G.J. and Dietvorst, A.G.J. (eds) (1995): 303–316.

Sinclair, M.T. and Stabler, M.J. (eds) (1991) *The Tourism Industry: An International Analysis*, Wallingford: CAB International.

Sinclair, M.T. and Vokes, R.W.A. (1993) 'The economics of tourism in Asia and the Pacific'. In Hitchcock, M. *et al.* (eds) (1993): 200–213.

Smith, R.A. (1991) 'Beach resorts: a model of development evolution', *Landscape and Urban Planning*, Vol. 21: 189–210.

Smith, R.V. and Mitchell, L.S. (1990) 'Geography and tourism: a review of selected literature'. In Cooper, C.P. (ed.) (1990): 50–66.

Smith, S.L.J. and Godbey, G.C. (1991) 'Leisure, recreation and tourism', *Annals of Tourism Research*, Vol. 18 No. 1: 85–100.

Smith, V.L. (1977) *Hosts and Guests: The Anthropology of Tourism*, Philadelphia: University of Pennsylvania Press.

Smith, V.L. and Eadington, W.R. (eds) (1994) *Tourism Alternatives: Potentials and Problems in the Development of Tourism*, London: John Wiley.

Soane, J. (1992) 'The origin, growth and transformation of maritime resorts since 1840', *Built Environment*, Vol. 18 No. 1: 12–26.

Steinecke, A. (1993) 'The historical development of tourism in Europe'. In Pompl, W. and Lavery, P. (eds) (1993): 3–12.

Teo, P. (1994) 'Assessing socio-cultural impacts: the case of Singapore', *Tourism Management*, Vol. 15 No. 2: 126–136.

Theobald, W. (ed) (1994) *Global Tourism: The Next Decade*, Oxford: Butterworth Heinemann.

—— (1994) 'The context, meaning and scope of tourism'. In Theobald, W. (ed.) (1994): 3–19.

Thomas, C. (1997) 'See your own country first: the geography of a railway landscape'. In Westland, E. (ed.) (1997): 107–128.

Towner, J. (1996) *An Historical Geography of Recreation and Tourism in the Western World, 1540–1940*, Chichester: John Wiley.

Tsartas, P. (1992) 'Socio-economic impacts of tourism on two Greek islands', *Annals of Tourism Research*, Vol. 19 No. 3: 516–533.

Turner, L. and Ash, J. (1975) *The Golden Hordes: International Tourism and the Pleasure Periphery*, London: Constable.

Urry, J. (1990) *The Tourist Gaze: Leisure and Travel in Contemporary Societies*, London: Sage.

—— (1995) *Consuming Places*, London: Routledge.

Valenzuela, M. (1988) 'Spain: the phenomenon of mass tourism'. In Williams, A. and Shaw, G. (eds) (1988): 40–60.

Voase, R. (1995) *Tourism: The Human Perspective*, London: Hodder and Stoughton.

Walton, J. (1993) 'Tourism and economic development in ASEAN'. In Hitchcock, M. *et al.* (eds) (1993): 214–233.

Walton, J.K. (1981) 'The demand for working-class seaside holidays in Victorian England', *Economic History Review*, Vol. 34 No. 2: 249–265.

Walvin, J. (1978) *Beside the Seaside: A Social History of the Popular Seaside*, London: Allen Lane.

Ward, C. and Hardy, D. (1986) *Goodnight Campers! The History of the British Holiday Camp*, London: Mansell.

West Country Tourist Board (1991) *Spreading Success: A Regional Tourism Strategy for the West Country*, Exeter: The Board.

Westland, E. (ed.) (1997) *Cornwall: The Cultural Construction of Place*, Penzance: Patten Press.

Wheeller, B. (1994) 'Ecotourism: a ruse by any other name'. In Cooper, C.P. and Lockwood, A. (eds) (1994): 3–11.

Wild, C. (1994) 'Issues in ecotourism'. In Cooper, C.P. and Lockwood, A. (eds) (1994): 12–21.

Wilkinson, P.F. (1987) 'Tourism in small island nations: a fragile dependence', *Leisure Studies*, Vol. 6 No. 2: 127–146.

Williams, A.M. and Shaw, G. (eds) (1988) *Tourism and Economic Development: Western European Experiences*, London: Belhaven.

—— (1995) 'Tourism and regional development: polarization and new forms of production in the United Kingdom', *Tijdschrift voor Economische en Sociale Geografie*, Vol. 86 No. 1: 50–63.

Witt, S. (1991) 'Tourism in Cyprus: balancing the benefits and costs', *Tourism Management*, Vol. 12 No. 1: 37–46.

World Tourism Organization (1994) *Aviation and Tourism Policies*, London: Routledge.

—— (1994) *National and Regional Tourism Planning*, London: Routledge.

—— (1995) *Compendium of Tourist Statistics 1989–1993*, Madrid: WTO.

# Index